Transbounday Hydro-Governance

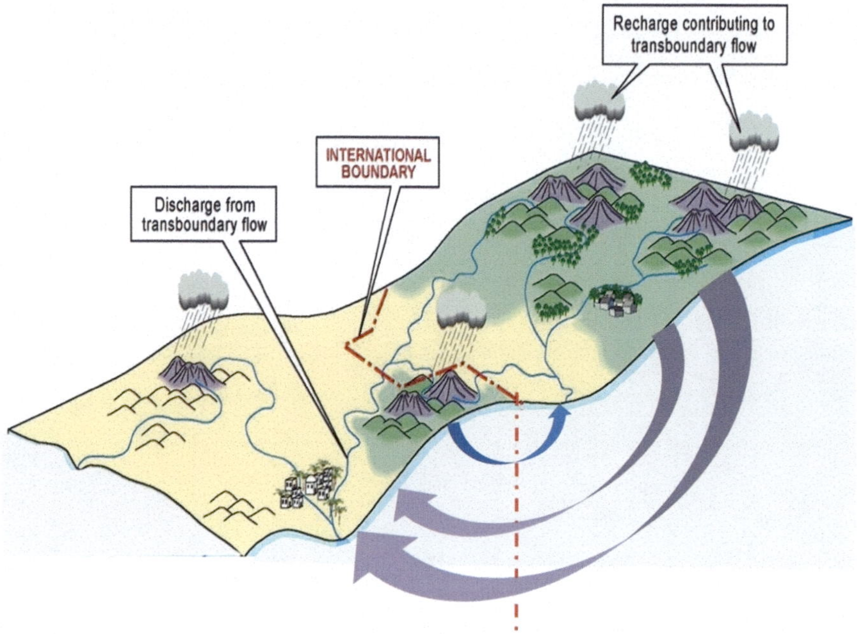

Transboundary hydrological and hydrogeological interactions (Adapted from UNESCO/ISARM 2001)

Jacques Ganoulis • Jean Fried

Transboundary Hydro-Governance

From Conflict to Shared Management

Springer

Jacques Ganoulis
UNESCO Chair/INWEB,
Civil Engineering Department
Aristotle University of Thessaloniki
Thessaloniki, Greece

Jean Fried
Department of Urban Planning and Public Policy
School of Social Ecology, University of
California
Irvine, California, USA

ISBN 978-3-030-08748-7 ISBN 978-3-319-78625-4 (eBook)
https://doi.org/10.1007/978-3-319-78625-4

Printed on acid-free paper

This Springer imprint is published by the registered company Springer International Publishing AG part of Springer Nature.
The registered company address is: Gewerbestrasse 11, 6330 Cham, Switzerland

Foreword

The UNESCO International Hydrological Programme (IHP) is the only intergovernmental programme of the United Nations system devoted entirely to water resources education, management and governance. The water crisis is primarily one of governance, where institutions lack the capacity to overcome conflicting approaches in the use and allocation of water from within one basin or aquifer system, both at national and transboundary level. With the shared use of freshwater resources being a key global challenge, many solutions to water management problems lie in improved water governance.

Driven by the growing recognition of the role of water governance in the context of increasing competition for limited water resources globally, the Intergovernmental Council of the UNESCO IHP launched the International Shared Aquifer Resources Management (ISARM) project in the year 2000, with the goal of developing wise practices and guidance tools for the shared management of groundwater resources and to contribute to the multifaceted efforts required for global water cooperation. A key result of the project was the publication by UNESCO of the first world map of 592 transboundary aquifers.

Despite progress made, in order to ensure effective and sustainable transboundary water governance and ensure the livelihoods of millions worldwide, innovative, practical and results-focused education and training is needed in order to prepare a new generation of water scientists, decision-makers and practitioners.

This book bridges the gap between theory and practice, considering the interactions of surface water and groundwater as they occur in the real world: governed by integrated transboundary hydro-governance. Jacques Ganoulis and Jean Fried have worked with the UNESCO IHP for many years and contributed extensively to the definition and development of transboundary groundwater education and training, especially within the framework of the corresponding UNESCO project. This book introduces what could be called "effective transboundary hydro-governance" and proposes practical instruments, tools and methodologies of its use within an analytical framework supported by multidisciplinary drivers.

In summary, this book comes at the very right moment, and will prove to be a pragmatic contribution to the study and understanding of transboundary water problems and their solutions, through the effective governance of shared water resources, supported by a robust and practical analytical framework. From policymakers and scientists to lawyers and field practitioners, this book will be both a guiding manual and a reference tool for all water managers in need of practical answers to real-world problems on the ground.

Groundwater Resources and Water for Human Section, Division of Water Sciences, International Hydrological Programme, UNESCO, Paris, France

Alice Aureli

Preface

A crucial challenge for humanity in years to come is how to ensure "***water security***", i.e. the necessary quantity and quality of water that man and nature need under any climatic conditions, including expected climate change and extreme droughts and floods. Failure to attend to such water security for all anthropogenic and environmental needs constitutes a ***major systemic risk*** for humanity in the near future. This is due to increasing global population, which means a higher demand for drinking water, economic growth with overconsumption of water, especially in agriculture, less precipitation in many parts of the world due to climate change and various forms of growing water pollution due to human activities. In this context, the key word for balancing water availability and water demand and achieving sustainable water security is ***hydro-governance***, i.e. an interactive process for managing at different levels and with different actors all kinds of water, including not only natural water resources but also new-water from wastewater after recycling and seawater desalination.

Hydro-governance becomes much more complex in transboundary water catchments, i.e. in river and lake basins shared by riparian countries. These catchments cover almost half of the world's land surface, with about 60% of global river flow and about 40% of the world's population. Moreover, in many parts of the planet, mainly in the Saharan Africa, in South America and in Central Asia, the population depends on transboundary groundwater resources located in aquifer systems shared by two or more countries. These facts illustrate the importance of the book's subject, which is ***transboundary hydro-governance***.

The book is written by two well-known scientists, both having an extended technical and policy expertise in the European Commission (EC) and various international organisations, such as UNESCO, GEF and UNEP. In hydro-governance, the main challenge is how to bridge the gap between science and technology on the one hand and law, policy and political decision-making on the other hand. Bridging this gap and involving institutions and actors from both the private and the public sectors and, in general, the civil society are the main factors for achieving an effective governance of shared water resources. What is new in this book is ***integrated transboundary hydro-governance***, which considers both surface water and groundwater as they really interact. In current scientific literature as well as in law and policy, groundwater is often missing for the benefit of surface water or treated separately from surface water resources.

The most important feature of the book is to distinguish between actual and ideal or ***effective transboundary hydro-governance*** and to provide methodologies, practical tools and examples for its implementation in the field. Published timely, this book addresses pragmatically worldwide problems, like climate change, significant floods and droughts especially in parts of the globe where every drop of water counts and in transboundary waters that represent a very important part of our precious freshwater resources.

Thessaloniki, Greece — Jacques Ganoulis
Irvine, CA, USA — Jean Fried

Contents

Part I
Main Aspects of Transboundary Waters

Chapter 1
Transboundary Waters

Water is the origin of everything.
Thales of Miletus, Greek philosopher, 624-546 BC

Abstract Transboundary waters are rivers, lakes and aquifers divided by political borders. Although the physical characteristics of internationally shared hydrological catchments and aquifers are independent from the socio-economic conditions of each riparian country, their scientific description and management may differ significantly from one country to another. This is the case, since countries and national institutions develop their own monitoring systems for collecting data and information, apply different scientific approaches due to historical and national socio-economic conditions and may provide different scientific interpretations of the same hydrological processes by national expert groups.

This chapter explains the importance of a shared approach to transboundary water resources assessment and management. After describing the typology of transboundary hydrological and hydrogeological systems, useful recommendations are formulated on how to deal with different national monitoring data and setting up common collaborative scientific approaches. Enabling cooperation between different groups of scientists, experts and policy makers from the riparian countries is important to proceed with the harmonisation of the national monitoring systems, obtain common interpretation of the data and set up shared river basin management plans.

Keywords International water resources · Transboundary groundwater · Socio-economics · Disputes · Collaboration

1.1 Transboundary Surface Water

1.1.1 The Water Cycle in Transboundary Areas

Our planet has plenty of water and therefore it is called the "blue planet". The total volume of water on Earth is estimated at 1360 million cubic kilometres or 1360×10^6

J. Ganoulis, J. Fried, *Transboundary Hydro-Governance*,
https://doi.org/10.1007/978-3-319-78625-4_1

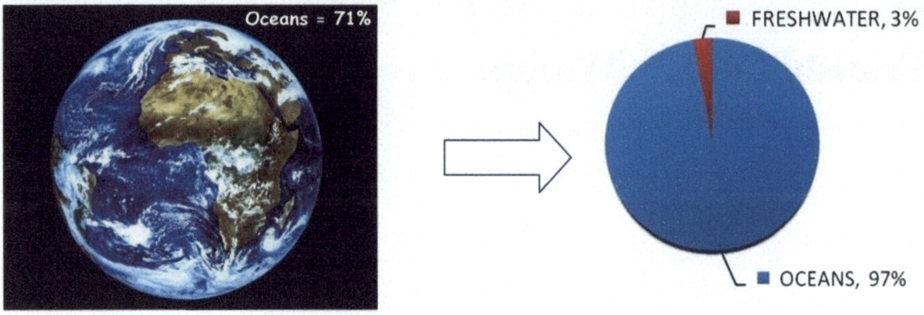

Fig. 1.1 Distribution of total water on Earth

Table 1.1 Distribution of freshwater on Earth (Gleick 1996 and USGS)

Component of freshwater	% Of the freshwater content (estimate)
Glaciers and permanent snow cover	68.7
Groundwater	30.1
Freshwater lakes	0.26
Rivers	0.006
Atmosphere	0.004
Biosphere	0.003

km^3 (Gleick 1996 and USGS). This number was derived from a long-term assessment of the average amount of water stored in the hydrosphere, i.e. the part of the Earth covered by water and ice, including the oceans, seas and groundwater, the water in the atmosphere and biosphere (all living organisms on Earth). The oceans and seas cover 71% of the surface of our planet and the volume of salt water in the seas and oceans represents 97% of total water, the remaining 3% being freshwater (Fig. 1.1).

Freshwater is distributed in different components (glaciers, rivers, lakes, groundwater, atmosphere and biosphere) as shown in Table 1.1 and Fig. 1.2. As it is shown, the greatest part (68.7%) of total freshwater is trapped in polar glaciers and ice sheets, and therefore it is not directly accessible for use. Only 0.3% of the freshwater on Earth is surface water, in the form of lakes (87%) and rivers (2%).

Water exists in three states: liquid, solid (ice and snow) and gas (water vapour). Due to the solar power, water is in permanent transformation from one state to another, and in constant motion between oceans, land, atmosphere and biosphere. As shown in Fig. 1.3, water in motion constitutes the *hydrological cycle* through the following hydrological processes, which take place in a permanently ongoing way (UNESCO glossary 2012):

Evaporation: emission of water vapour by a free surface at a temperature below boiling point

Transpiration: transfer of water vapour from vegetation to the atmosphere

Evapotranspiration: combination of evaporation and transpiration

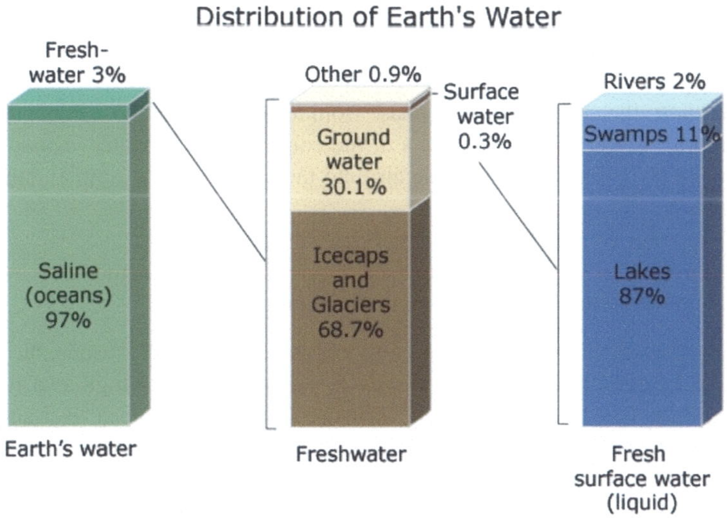

Fig. 1.2 Distribution of total water on Earth (Gleick 1996 and USGS)

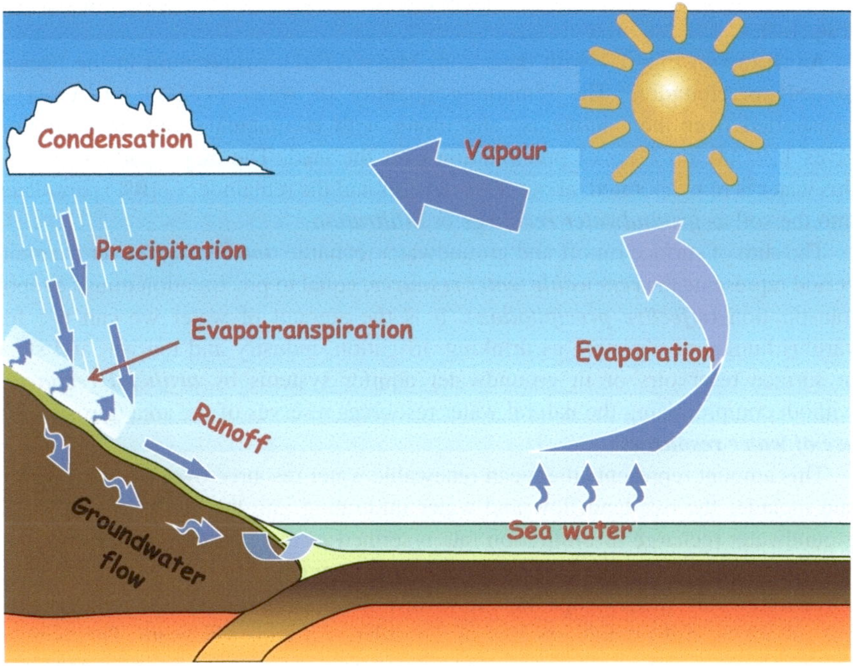

Fig. 1.3 The hydrological cycle

Interception: process by which precipitation is caught and held by vegetation (canopy and litter structures) that may then be lost by evaporation without reaching the ground
Condensation: the change in water phase from a vapour state into a liquid state
Precipitation: liquid or solid products of the condensation of water vapour falling from clouds or deposited from the air onto the ground. For example, rain, sleet, snow, hail
Runoff: that part of precipitation that appears in surface streams
Infiltration: flow of water through the soil surface into a porous medium
Groundwater flow: movement of water in an aquifer.

The hydrological cycle has the extraordinary property of sustainable cyclic movement by combining the solar energy to water physical change as a function of temperature. Due to solar power the saline sea water and water in river and lakes and that intercepted by terrestrial vegetation are evaporated into the atmosphere. Then, after condensation in the clouds, it returns to the land and to oceans and seas in various liquid forms, such as water, snow or hail, depending on the temperature.

The total hydrological cycle and the so-called "water budget" or "water balance" can be considered at various scales such as global, regional and local. At global scale it is useful to distinguish between the terrestrial or land and the marine water cycle (Fig. 1.4).

As shown in Fig. 1.4 with data from Musy (2001), evaporation in the oceans exceeds precipitation. The remaining quantity of water (47,000 km^3/year) is transported over the continents, and along with continental evapo-transpiration gives 119,000 km^3/year of precipitation over the land. The major part (~60%) of this water is re-evaporated, about 30% runs off and the remainder (~10%) percolates into the soil as ***groundwater recharge*** or ***infiltration***.

The sum of surface runoff and groundwater recharge (***total runoff***) over a certain period represents the ***renewable water resource,*** equal to precipitation minus evapo-transpiration (***effective precipitation***). It is the amount of water we can use for various human needs, such as drinking, irrigation, industry and tourism and store in surface reservoirs or in groundwater aquifer systems by ***artificial recharge***, without compromising the natural water resources reserves of the area (***sustainable use of water resources***).

This amount represents the mean renewable water resource over a certain period and includes the overland flow and water infiltrating into the soil. The renewable groundwater recharge or infiltration rate is termed ***meteoric groundwater*** and is of great importance for groundwater resources development. For shallow aquifers with a free water table, the replenishment time for water cycling is of the order of a year or less, although in deep aquifers it is of the order of a thousand years. The average replenishment rate of an aquifer from natural and artificial recharge is generally considered as the ***safe yield***.

For the development of water resources, it is important to know the spatial distribution of various components of the hydrological cycle. For groundwater resources planning, it is necessary to have geological, hydrogeological and

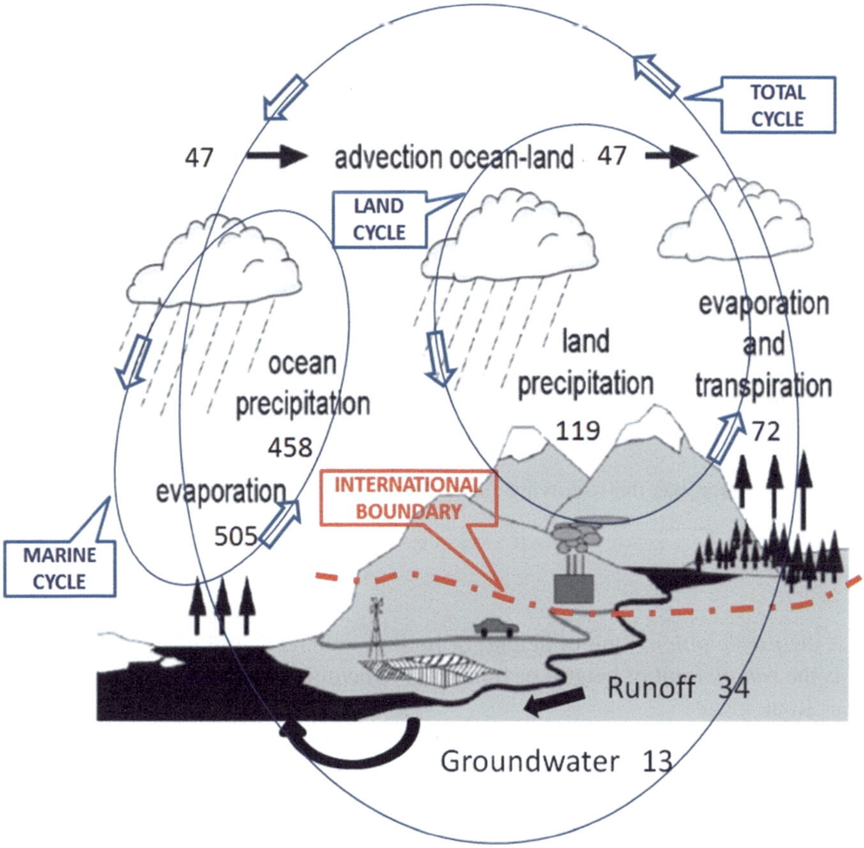

Fig. 1.4 Global, land and marine hydrological cycles; numbers in thousands km^3/year (adapted from Quante 2007)

hydrochemical information and field data, allowing the evaluation of the groundwater balance.

For the EU region, the mean annual volume of precipitation water is estimated at 1375 km^3/year (97 cm/year) and the efficient precipitation at 678 km^3/year (48 cm/year) (Bodelle and Margat 1980). Main hydrological data for some European Union countries are summarised in Table 1.2.

For water resources management in a specific hydrological area or at the catchment scale, it is necessary to quantify the available water resources for a given time scale. The water balance or the water budget of a region is the quantification of the individual components of the water cycle during a certain time interval. The water budget is mathematically computed with the continuity equation, which expresses the balance between the inflows, outflows and change of storage in any water body/hydrologic region over a period.

Table 1.2 Mean annual precipitation and flows (surface runoff plus infiltration) in some EU countries (Bodelle and Margat 1980)

Country	Precipitation		Efficient precipitation (surface runoff plus infiltration)	
	Mean annual volume	Mean annual height		
	$\times 10^9$ m^3/year	cm/year	$\times 10^9$ m^3/year	cm/year
France	440	80	170	30
Greece	112	85	55	42
Italy	300	100	165	55
Spain	330	66	110	22
Germany	200	80	86	35
United Kingdom	250	100	160	65
Belgium	250	85	110	36

This equation has the following form (Musy 2001):

$$P - R - G - E - T = \Delta S \tag{1.1}$$

where

P is the *precipitation*, [unit of height] or [unit of volume/unit of time]
R is the *runoff*, [unit of height] or [unit of volume/unit of time]
R = Rout—Rin
Rout = runoff as outflow from the water body/hydrologic region
Rin = runoff as influx into the water body/hydrologic region
G is the *groundwater flow*, [unit of height] or [unit of volume/unit of time]
G = Gout—Gin
Gout = groundwater as outflow from the water body/hydrologic region
Gin = groundwater as influx into the water body/hydrologic region
E = *evaporation*, [unit of height] or [unit of volume/unit of time]
T = *transpiration*, [unit of height] or [unit of volume/unit of time]
ΔS = change in storage, [unit of height] or [unit of volume/unit of time]

Equation 1.1 is the basic equation of hydrology. In practice, it can be successfully applied for local studies when the various hydrologic terms can be properly measured or estimated. Nevertheless, estimation is usually rough on a global scale.

1.1.1.1 Examples of Application

Annual Water Budget of the Earth (see Fig. 1.5)

Every year approximately 577,000 km^3 of water are transported through the hydrosphere in form of precipitation. The annual global water budget is shown in Fig. 1.5.

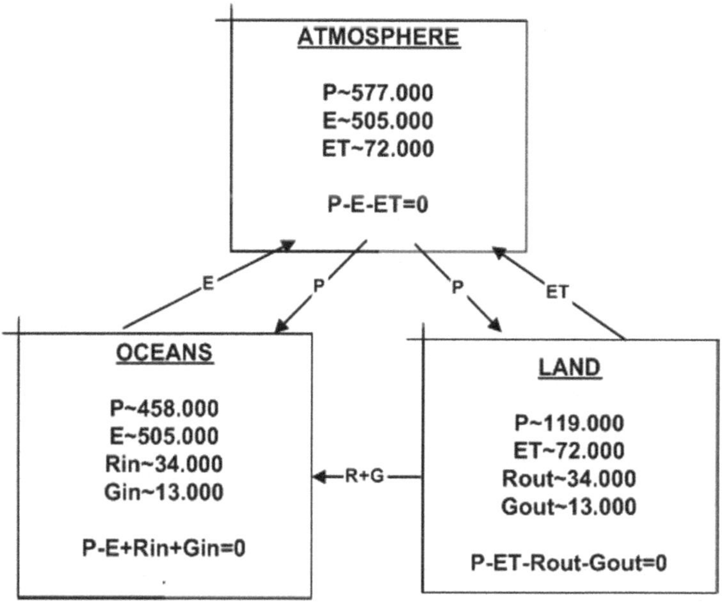

Fig. 1.5 Global annual water balance (Musy 2001)

Annual Water Budget of Switzerland (Table 1.3)

Table 1.3 Annual water budget of Switzerland (Musy 2001)

Precipitation			1546 mm/year
Total runoff	Runoff	978 mm/year	1296 mm/year
	Influx into Switzerland	318 mm/year	
Evaporation			484 mm/year

Annual Water Budget of Romania (Table 1.4)

Table 1.4 Annual water budget of Romania (National Institute of Meteorology and Hydrology, Regional Office, Timisoara)

Precipitation	850 mm/year
Runoff	300 mm/year
Evaporation	550 mm/year

Annual Water Budget of Bulgaria (Table 1.5)

Table 1.5 Annual water budget of Bulgaria (Geography of Bulgaria, monograph, Bulgarian Academy of Sciences, 1989)

Precipitation	690 mm/year
Runoff	176 mm/year
Evaporation	514 mm/year

Annual Water Budget of the Ukraine (Table 1.6)

Table 1.6 Annual water budget of the Ukraine

Precipitation	625 mm/year
Runoff	86.8 mm/year
Evaporation	538 mm/year

Figure 1.6 shows an example of a cross border surface relief divided by an international boundary. In this transboundary area, complex hydrological interactions may take place between the two countries. This is the case where one country is in higher altitude upstream and thus, precipitation in the mountains is contributing to transboundary river flow downstream to the other country. This means that water recharge in the upstream country may generate discharge of surface water in the downstream one.

Groundwater interactions between two countries along the international boundary may be more complicated. As shown in Fig. 1.6, in deep aquifer layers, groundwater is directed to the downstream country although near the boundary, in shallower depth, groundwater may move in the opposite direction, i.e. from the downstream to the upstream country. Furthermore, there are well known case studies, which illustrate that pumping wells near the border from one country may induce negative effects to the other riparian country, such as reducing the water volume of a lake located in the other part of the boundary or contributing to sea water intrusion in coastal aquifers located in the second country.

Difficulties in developing national water management plans in transboundary areas are mainly due to the following natural, socio-economic and political reasons:

Natural Reasons

- *Hydrological and hydrogeological interactions* across the international boundary are complex and difficult to be assessed.
- *Hydro-meteorological and hydrological* data collected at national scale need to be harmonised if not supported by a common monitoring system.
- Hydrological *upstream-downstream asymmetry* and different hydro-geological interactions between the riparian countries generate the main problem of sharing the amount of water flowing from one country to another.

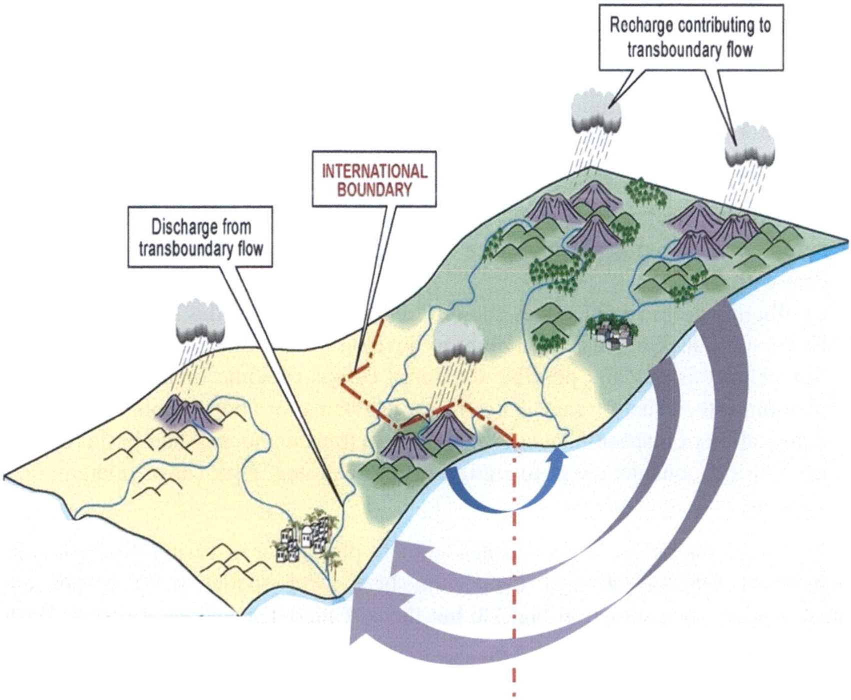

Fig. 1.6 Transboundary hydrological and hydrogeological interactions (adapted from UNESCO/ISARM 2001)

- *Water quality* issues may induce severe problems between riparian countries, mainly when upstream countries export polluted water to downstream located countries.
- *Hydrological extremes,* such as floods and droughts originated in one riparian country may cause harm to another located downstream.

Socio-Economic Reasons

- *Different socio-economic conditions* between riparian countries mean different needs for water in transboundary river catchments and different allocation between economic sectors, such as agriculture, tourism, industry and environmental protection of shared water. For example, a poor country located upstream may need to use more water, hence constructing dams that reduce the water availability downstream, which may be detrimental to a riparian country located downstream.
- *Different socio-economic priorities and related targets* set by riparian countries may lead to incompatible water management plans in transboundary rivers and lakes.

- Different *national water laws and related policies* may create problems for managing water use and water quality protection in transboundary water basins. For example, upstream countries may neglect wastewater treatment plants, which may cause water pollution problems to countries located downstream.

Political Reasons

- Lack of political willingness: because transboundary water related issues are complex and need application of a long-term policy, politicians are reluctant to deal with them.
- Conflicting political objectives and different political priorities between countries may lead to lack of trust and failure to develop mutual interests.
- Sovereignty issues and possible territorial claims of some riparian countries in relation to their history may cause major problems for cooperation, and
- Application of international law and policy in transboundary water basins depend on political considerations for ratifying international, regional or bilateral conventions and agreements.

For developing effective water management plans after assessing different components of the hydrological cycle, the best local spatial scale is not any transboundary area shown in Fig. 1.6 but the so-called Transboundary River Basin (TRB, Fig. 1.7).

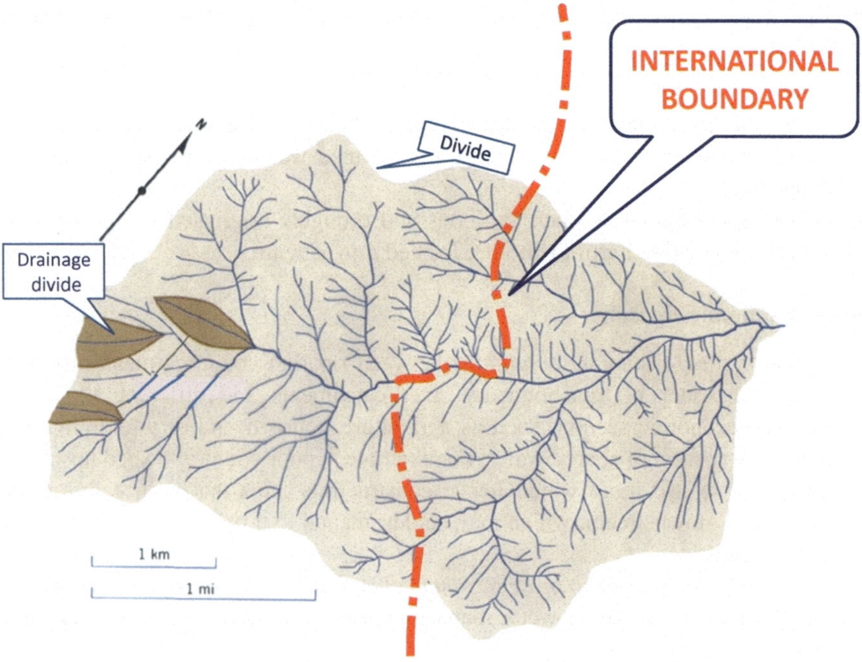

Fig. 1.7 Transboundary River Basin (TRB) defined by its water divide

A river basin is the area enclosed by the ***watershed boundary***, i.e. the ***divide*** of surface waters (Fig. 1.7). When a drop of rain water falls near this line, it flows either inside or outside the basin. Since no water is entering or exiting the basin along the watershed border, the water balance in the basin is simply expressed by the effective precipitation, which is equal to the total runoff. As shown in Fig. 1.7, due to different effluent streams inside the main river basin, many secondary river basins and drainage divide areas are defined. This involves the need for a lot of measurement stations and makes calculation of the hydrological components very difficult and almost approximate.

When surface waters, like rivers and lakes are divided by an international boundary, the terms "*Transboundary Waters*", "*Transboundary River Basin*" and "*Transboundary Lake Basin*" are used as synonymous with "*Internationally Shared Waters*". This is in accordance with the terminology used by UNESCO in its international hydrological initiatives, such as the UNESCO/ISARM (Internationally Shared Transboundary Aquifer Resources Management) and the UNESCO/PC-CP (Potential Conflict-Cooperation Potential) programmes. This terminology is considered as a better choice than other similar expressions such as "*International Waters*", or "*Multinational Waters*" or "*Regional Waters*", because it avoids misunderstandings due to political sensitivities over national sovereignty on regions located near political borders.

Figure 1.8 shows two characteristic examples of transboundary surface waters, shared between two and three countries located in South Eastern Europe (the Balkans).

In contrast with the lake basins, river basins terminate with an outflow into a sea or an ocean. At a regional scale, a set of river basins may have an outflow into a regional sea. The Mediterranean basin shown in Fig. 1.9 covers all the drainage areas of rivers flowing into the Mediterranean Sea. If we add on the map the political boundaries, we can define individual transboundary river basins, such as the Mesta/Nestos River Basin shared between Bulgaria and Greece.

It is interesting also to see that coastal national waters, defined by the EU/WFD (Water Framework Directive) as the zone within a distance of 1 mile from the coast, are also "***internationally shared***" between the Mediterranean countries. This means that due to coastal circulation and various marine currents in the coastal zone, pollution from one country can affect other coastal Mediterranean countries. These interactions should be taken into consideration for preserving fisheries, the marine biodiversity and other ecosystems functions. This explains the importance of global Interstate Agreements and Mediterranean Treaties, like the ***Union for the Mediterranean*** and the ***Barcelona Convention*** for protecting and preserving the good status of the Mediterranean. The same applies to all regional and open seas around the globe.

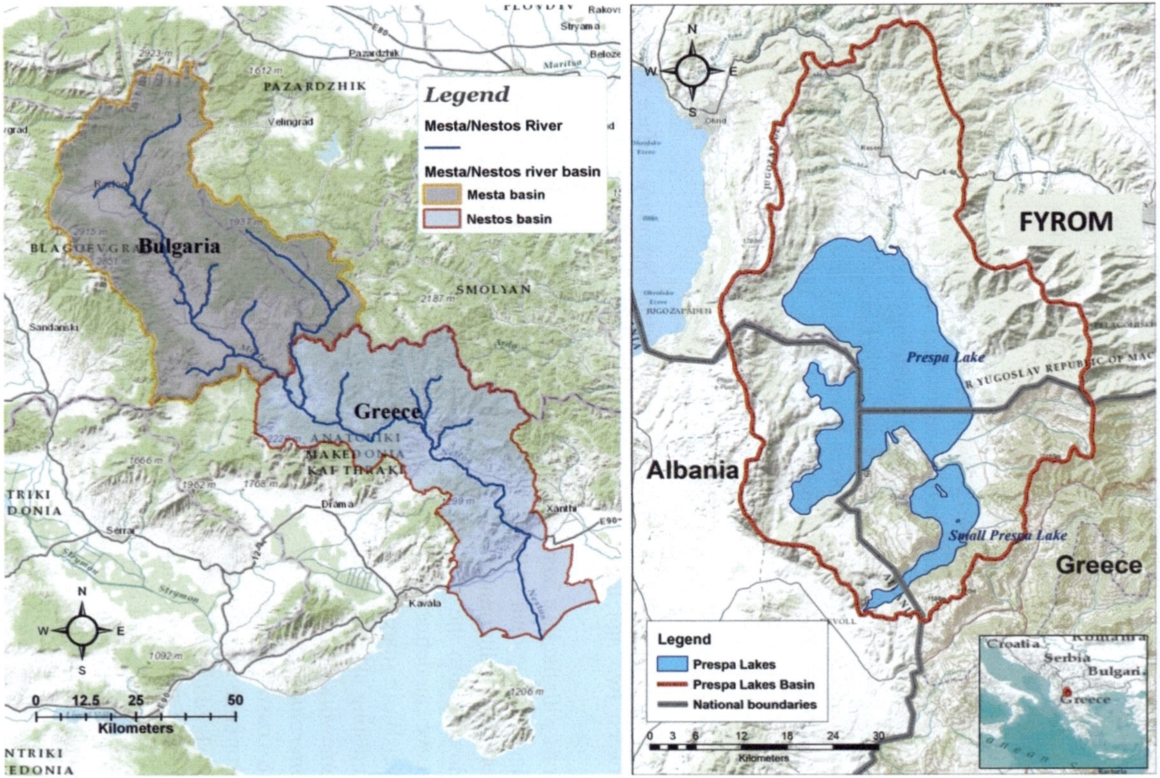

(a) (b)

Fig. 1.8 The TRB Mesta/Nestos between Bulgaria and Greece (**a**) and the Transboundary Prespa Lakes Basin between Albania, the former Yugoslav Republic of Macedonia (FYROM) and Greece (**b**)

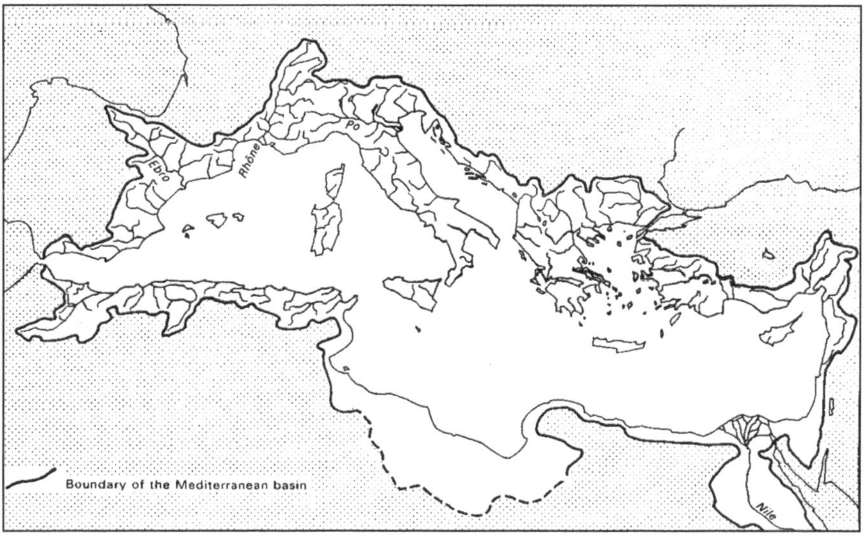

Fig. 1.9 The Mediterranean basin

1.1.2 Transboundary River and Lake Basins

Transboundary surface water resources on earth are a substantial part of the total amount of freshwater available worldwide. According to the UN World Water Development Report (UN WWDR3 2009) demographics and the increasing water over-consumption are the most important pressures on water. The world's population is growing by about 80 million people a year, implying increasing freshwater demand of about 64 billion cubic metres a year. Agriculture is the largest consumer of freshwater by far—about 70% of all freshwater withdrawals go to irrigated agriculture. The global renewable water availability is estimated around 13,500 km^3 per year, which gives 2300 m^3 per person per year. This is approximately 37% less than the amount in 1970.

The global increase of population together with the steady socio-economic development, especially of emerging economies, and the subsequent increase in water demand combined with the acceleration of water pollution from various point and diffuse sources mean that transboundary water resources located both on surface (rivers and lakes) and groundwater aquifers are important sources of water for different uses at global and regional scales, and form a significant part of the precious available water on earth.

About 60% of the global river flow lies within transboundary river basins (Wolf et al. 1999) the surface area of which amount to almost half of the world's land surface, as shown in Fig. 1.10.

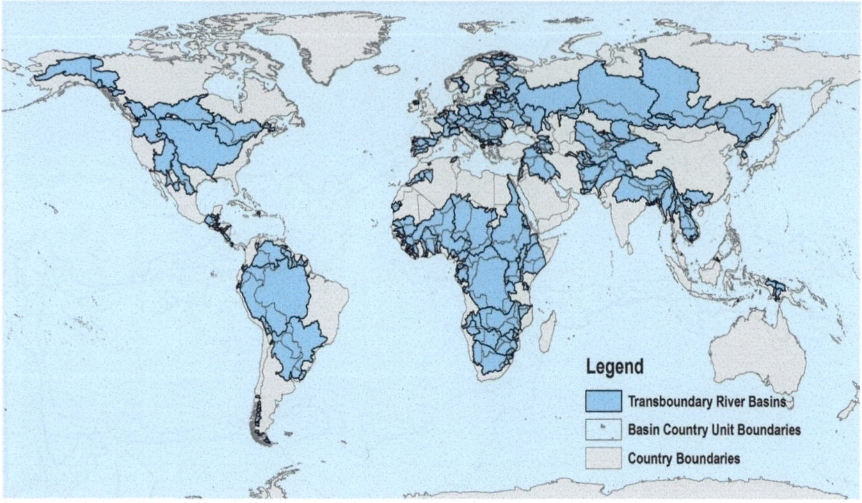

Fig. 1.10 Transboundary river basins worldwide (TWAP 2016)

Table 1.7 Number of transboundary river basins by continent and in % of total continent's surface (TWAP 2016)

Continent	1978 UNDESA	1999 OSU	2002 OSU	2012 OSU	2016 TWAP	% Of total continent's surface
Africa	57	60	59	64	63	62
Asia	40	53	57	60	68	39
Europe	48	71	69	68	71	54
North and Central America	33	39	40	46	45	35
South America	36	38	38	38	39	60
Total	214	261	263	276	286	45

The significance of transboundary waters can be seen from the following figures:

40% of the world's population lives within these watersheds
45% of the total land surface of our planet lies in this area
286 major internationally shared basins are reported

The actual number of transboundary basins may change not only because new political states emerge or in some cases, for example Germany, become unified, but also because cartographic methods improve.

As shown in Table 1.7, the number of transboundary basins varies over time in all continents. Due to the collapse of the Former Republic of Yugoslavia and the Soviet Union, the number of transboundary basins in Europe increased from 48 in 1978 to 69 in 2002; in South East Europe (SEE) alone this meant that the number of transboundary basis almost doubled. Globally (Table 1.7) in 2002 there were 286 transboundary basins listed, compared to 214 in 1978, 261 in 1999 and 276 in 2012.

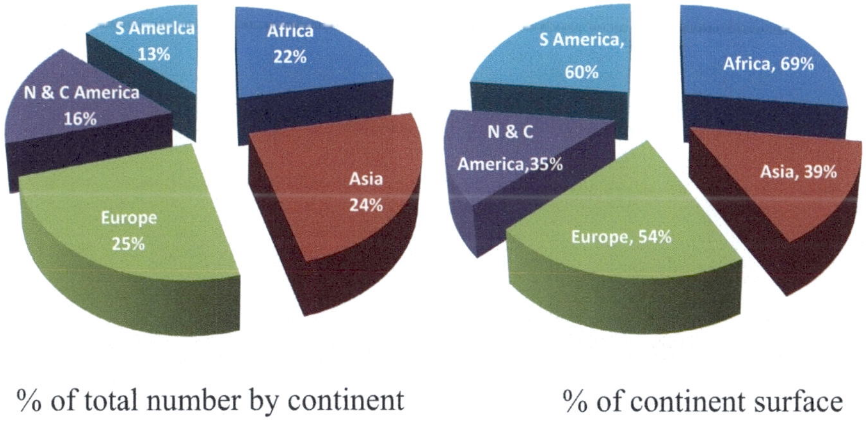

Fig. 1.11 World's transboundary river basins by continent (data from TWAP 2016)

Approximately one third of the transboundary basins (UN-Water 2018) are shared by more than two countries:

148 countries have territory within transboundary river basins
21 countries lie entirely within one transboundary river basin
39 countries have more than 95% of their territory within one or more transboundary basins
19 basins involve five or more different countries

The distribution of transboundary basins per continent by number and as a percentage of the continent's surface is given in Fig. 1.11: Europe has the greatest number of internationally shared basins (25%), while Africa's transboundary basins cover the greatest part of the continent (69%).

1.1.3 Typology of Transboundary River Basins

Different types of transboundary river basins can be classified depending on the position of the international boundary. As shown in Fig. 1.12, we may have:

1. Rivers that serve as borders between states, like the lower course of the Danube River, which serves as the border between Bulgaria and Romania (Fig. 1.12a, c); and
2. An international boundary that crosses the river at a given location and divides the river catchment into two parts, the upstream and the downstream (Fig. 1.12b, d). In this case, the river crosses the international boundary between the two states. This is the case of the border between Hungary and Serbia that is crossed by the Danube and Tisza Rivers or the border between Greece and the Former Yugoslav Republic of Macedonia (FYROM) crossed by the Vardar/Axios) River near the city of Gevgelia/Evzoni.

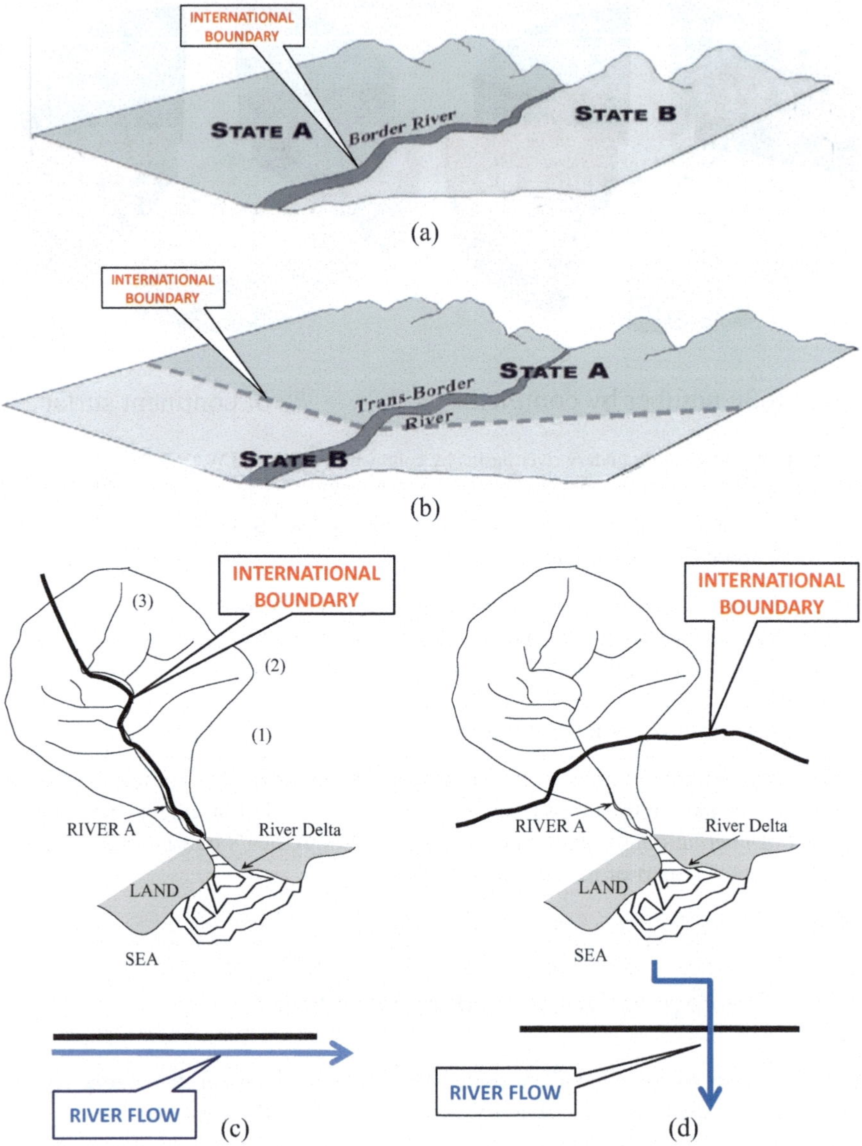

Fig. 1.12 Schematic representation of two types of transboundary river basins with international boundary: (**a**) & (**c**) following the river and (**b**) & (**d**) crossing the river

We may also have a mixed situation, where the international boundary follows the river over a certain distance and then crosses the river.

How the interstate borders follow and/or cross international rivers, and how they divide rivers and river basins, will determine what type of water resources problems exist or will likely arise and need bilateral or multilateral interstate solutions. For transboundary waters, many international agreements for solving various types of interstate water resources problems are available. The most important international treaty is the United Nations Economic Commission for Europe (UNECE) Convention.

1.2 Transboundary Aquifers

1.2.1 Importance of Groundwater

Fresh groundwater is often called the Earth's "hidden treasure" or humanity's "liquid gold" but despite being vital for sustaining life and socio-economic development at local and regional scales, it remains a poorly understood resource both by the large public as well as by policy makers. Estimations in the literature of the quantity of available groundwater resources at a global level vary by some orders of magnitude. Despite this fact, it is generally accepted that among all other sources of freshwater, groundwater is the most important, except for the vast quantities of freshwater blocked in icecaps and glaciers. Some facts that underline the importance of groundwater are the following:

- it is ubiquitous and represents more than 98% of all liquid freshwater on Earth
- about half the world's population drinks groundwater every day
- groundwater contributes to more than half of the world's irrigated agricultural production
- many industries prefer groundwater as their principal source of water.

Figure 1.13 shows that although groundwater is just the 1.7% of the Earth's total water, however it constitutes an important 30.1% of the Earth's total freshwater resources (USGS 2017). Of the total groundwater available, 54% is saline and the remaining 46% is fresh groundwater (Fig. 1.13).

It is no coincidence that many big cities and human settlements around the world have been developed near springs and aquifers, where groundwater emerges, especially in arid and semi-arid climate zones, where surface water is limited. Groundwater also feeds streams, supports ecosystems and shapes our environment. It forms an unrivalled water buffer.

By looking at the global map of groundwater resources (Fig. 1.14, prepared by WHYMAP UNESCO), we realise that the world under our feet is very complex and difficult to understand. For example:

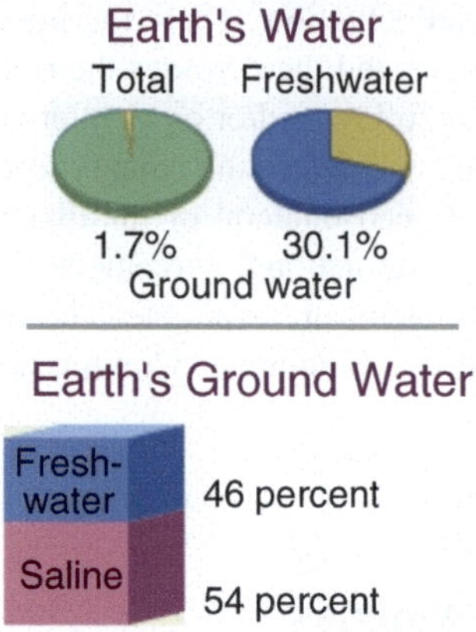

Fig. 1.13 Groundwater as a proportion of Earth's total water and Earth's freshwater (adapted from USGS 2017)

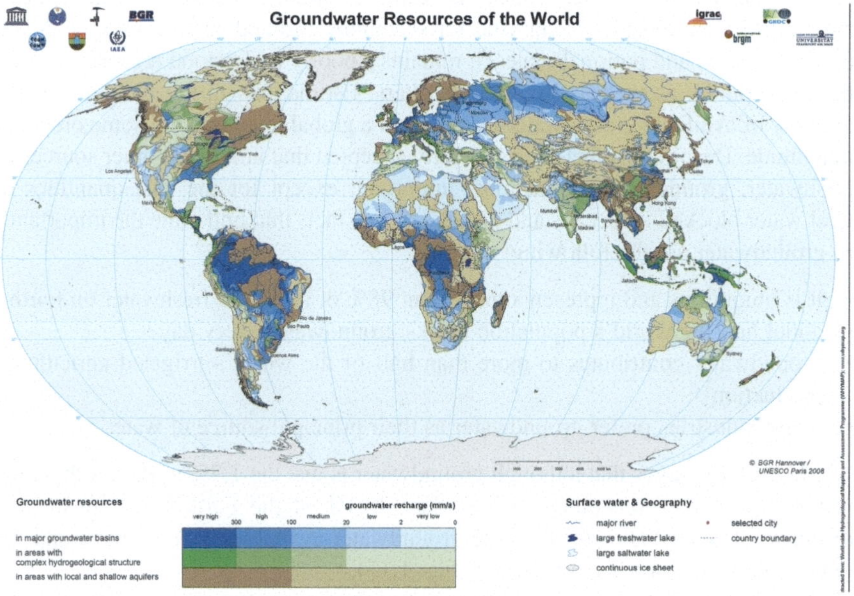

Fig. 1.14 Groundwater resources of the world (WHYMAP 2008)

- the African continent is not entirely made of sand and stones/dust as most commonly believed. Major aquifers in many sub-Saharan regions are shared by many countries and tensions in particular regions can be read in terms of sharing their transboundary water resources; and
- if water is the future "blue gold", then the map shows possible conflicts for sharing it. Pressures from demographic growth, climate change and other drivers

are putting a high toll on groundwater resources. Such pressures will prove, even more strongly, that sooner or later "we all live downstream" and that cooperation—not conflict—is really our last resort for survival.

Moreover, aquifers know no boundaries as any other natural resource, therefore their sound and concerted management is of utmost importance for any peaceful and sustainable development we envisage and this calls for increased diplomatic efforts globally. What groundwater governance can do to reverse negative trends? How can the media be better informed to convey facts and priorities related to groundwater management? How can this resource be better protected for the present and future generations? These questions are all crucial, to be adequately answered right now.

1.2.2 Transboundary Aquifers Around the World

Most of the global freshwater on Earth is found in aquifers, many of them transboundary. Characteristics of the aquifers are different from those of surface water bodies. An aquifer system consists of a series of two or more aquifers that are hydraulically connected. A transboundary aquifer system is an aquifer, parts of which are situated in different states. Currently more than 400 transboundary aquifers have been identified crossing the political boundaries of over 145 nations (Fig. 1.15).

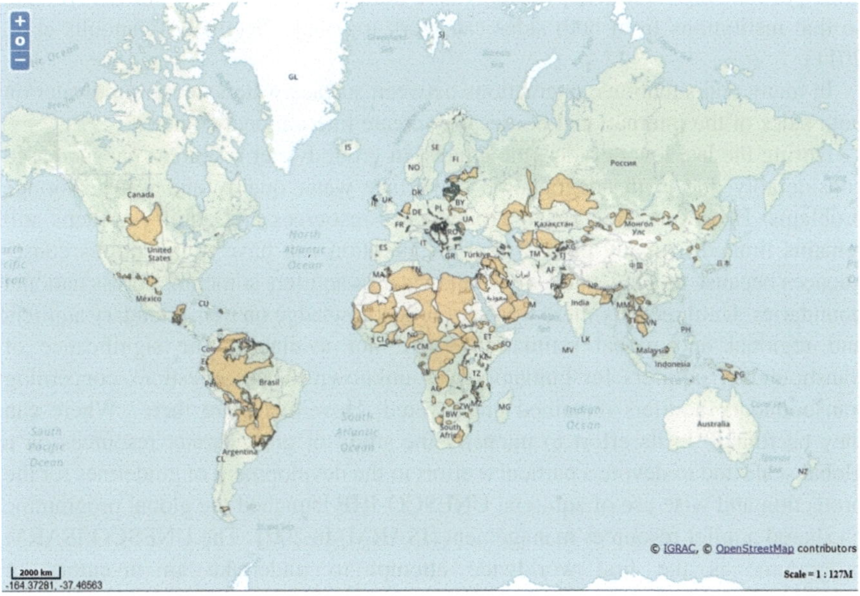

Fig. 1.15 Transboundary groundwater aquifer systems (IGRAC 2017)

Apart from their significance in terms of area and conflict potential, it should be noted that these basins:

- Cover 45% of the land surface of the Earth;
- Affect 40% of the world's population;
- Account for approximately 80% of global river flow.

In the case of internationally shared surface waters, much progress has been made on how to determine what type of water resources problems need or will need to be resolved by bilateral or multilateral interstate agreements. A large number of such international agreements (more than 200 during the last 50 years) have been signed to resolve various types of interstate surface water resources problems. An international convention has already been agreed upon for the monitoring and assessment of transboundary rivers and lakes, the UNECE Convention on the Protection and Use of Transboundary Watercourses and International Lakes, 1992.

The situation is quite different in the case of internationally shared groundwater aquifer resources: while the UNECE Convention has been signed in 1992, no such international binding treaty yet exists for transboundary aquifers. Difficulties arise in scientific and technical matters (groundwater monitoring, data interpretation and modelling), and due to the lack of political willingness for cooperation and the weakness of the institutions involved. Major difficulties arise in designing groundwater development plans, because groundwater flow and groundwater quality are subject to several types of uncertainties and are affected by these to a much greater degree than surface water. These uncertainties are related to the high variability in space and time of the hydrogeological, chemical and biological processes. The principal challenge is to set up a cooperative framework between countries involved, so that institutions from both sides can work together effectively (Ganoulis et al. 2011).

In many real situations, interactions between surface waters and groundwater on both sides of the international border may create international disputes.

During the last few decades, the interest in groundwater resources has increased considerably due to the degradation of surface water quality and water shortage problems. However, attention to groundwater resources and aquifer systems still remains limited in many places. Lack of attention can have more serious consequences because the flow paths of groundwater in aquifers sometimes cross national boundaries. Until recently there was very little knowledge on transboundary aquifers and regional and global estimations were not available. The significance of transboundary aquifers for humanity was unknown. Many questions concerning transboundary aquifers remained unanswered: How many are there? Where can they be found? In its effort to intensify the study of groundwater resources on a global scale and to devote a particular effort to the development of guidelines for the protection and wise use of aquifers, UNESCO-IHP launched the global programme on shared aquifer resources management (ISARM) in 2001. The UNESCO ISARM programme is the first worldwide attempt to undertake an inventory of transboundary aquifers across all continents and to set up criteria for the sustainable management of these systems. ISARM has a multidisciplinary focus at its core, considering scientific-hydrogeological approaches, legal aspects, socio-economic issues, institutional considerations and environmental protection issues.

Among other accomplishments, the ISARM network prepared the Hydrogeological Map of the World (WHYMAP) and published the first worldwide map and atlas of transboundary aquifers. It has, as a result, successfully contributed to better defining the characteristics of the world's groundwater resources on a global scale, and to the formulation of strategies for the protection and sustainable management of shared aquifer resources. The ISARM inventory showed that transboundary aquifer systems can extend over thousands of square kilometres, contain substantial amounts of water, often playing a crucial social and ecological role, and that the large majority of these systems are not directly connected with any international river basins.

Among the efforts to recognize the legal significance of Transboundary Aquifers, a special mention should be made of the UN General Assembly Resolution on the "Law of Transboundary Aquifers". The UN International Law Commission (ILC), with the technical support of UNESCO-IHP, embarked on the codification of the law of transboundary aquifers in 2002 to provide a legal regime for the sustainable management of aquifers. A UN General Assembly Resolution (A/RES/63/124) was adopted in December 2008, including the set of 19 draft articles prepared by the UNILC as an annex and encouraging States sharing an aquifer to consider them when entering an agreement for the proper management of the underground resource. At its sixty-sixth session in December 2011, the UN General Assembly reaffirmed the importance of transboundary aquifers and the related draft articles. In a new Resolution (A/RES/66/104), states are further urged to conclude appropriate bilateral or regional arrangements for managing their transboundary aquifers, and UNESCO-IHP is encouraged to continue providing its scientific and technical support to the states concerned. The draft Articles provide a comprehensive understanding of transboundary aquifers and their adequate management that includes procedures for data exchange, monitoring and cooperation. The articles also consider the existing disparities in capacity and knowledge between countries and the need for international technical cooperation. The articles include requirements on cooperation, including scientific and technical cooperation for developing countries, and regular exchange of data and information between the countries in whose territory the aquifer is located, as well as provisions concerning monitoring and prevention of contamination, with special attention to groundwater dependent ecosystems. The first example of the application of the draft Articles (annexed to both resolutions) is the agreement on the Guarani Aquifer signed by the four countries concerned (Argentina, Brazil, Uruguay and Paraguay) in 2009. This is the first international agreement that explicitly refers to the draft Articles as inspiring source/principles for its formulation and implementation.

Recognizing the importance of conjunctive management of water resources and the inextricable linkages between groundwater and surface water and ecosystems, a new synergy was created at the end of 2012, when, at the Meeting of the Parties of the UNECE Water Convention, it was unanimously agreed that the three major international instruments providing guidelines for water management, i.e. the

Convention on the Protection and Use of Transboundary Watercourses and International Lakes, 1992, whose Secretariat is located within the UNECE, the 1997 UN Convention on the Law of the Non-Navigational Uses of International Watercourses and the Draft Articles of the Law of Transboundary Aquifers, would join forces and complement each other.

1.2.3 Typology of Transboundary Aquifer Systems

Water infiltrating into the soil circulates through various geological formations. Depending on the boundary conditions (impermeable or semi-permeable layers of soil, atmospheric pressure, rivers and lakes), groundwater forms various types of subsurface reservoirs, called aquifers. These are extensive permeable rock formations in which water partially accumulates and through which water partially flows. Figure 1.16 gives an overview of different types of groundwater aquifers in various geological formations. According to their geological formation characteristics, aquifers may be classified in three main groups:

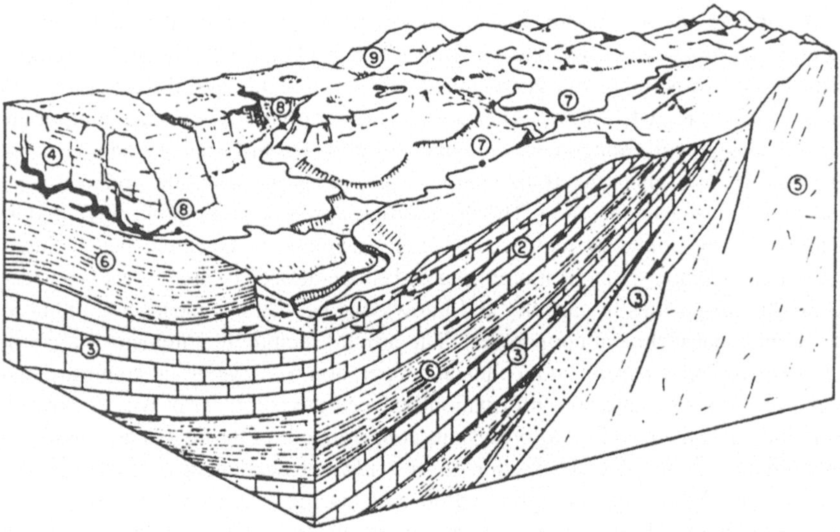

1 Alluvial aquifer in connection with a river
2 Regional aquifer with free surface in limestone
3 Regional aquifer between impermeable layers
4 Karstic aquifer
5 Crystalline fractured rock
6 Sedimentary impervious layer
7 Spring from phreatic aquifer
8 Spring from karstic aquifer
9 Leakage to a river

Fig. 1.16 Groundwater in various geological formations (Bodelle and Margat 1980)

1. Alluvial and sedimentary aquifers.
2. Limestone and karstic aquifers.
3. Crystalline fractured rock aquifers.

1. Alluvial and sedimentary aquifers: This category of aquifers is characterized by successive layers of different hydrogeological properties: permeable, semi-permeable or impermeable. The water circulates in the successive layers, which consist mostly of gravel, sand, clay and silt. Phreatic, confined or semi-confined (leaky) aquifers are formed.
2. Limestone and karstic aquifers: Solution processes caused by acidified rainwater increase the permeability of limestones and dolostones forming secondary aquifers. Karstic phenomena are extreme cases of such processes, creating subterranean fractures and water conduits of high permeability. In karstic regions surface runoff is almost nil and large volumes of groundwater can be found at various depths.
3. Crystalline fractured rock aquifers: The importance of groundwater resources in these rocks depends on two factors (1) the rate of fracturing and (2) the chemical weathering of the surface layer, through which precipitation water percolates into the rock. This geological formation is divided in several blocks by secondary and primary fractures.

Transboundary aquifers may be classified as shown in Fig. 1.17, based on the position of the international boundary and the aquifer type: alluvial, karstic, crystalline, other (Fried and Ganoulis 2016).

- Type (a): a river is the international border in connexion with an unconfined transboundary aquifer. Little transboundary groundwater flow occurs;
- Type (b): the state border intersects the river and the alluvial aquifer. Significant groundwater discharge occurs from the upstream to the downstream country;
- Type (c): an unconfined aquifer that flows across an international border and is hydraulically linked to a river that flows completely within the territory of one state.;
- Type (d): an unconfined aquifer that is completely within the territory of one state but is linked hydraulically to a river flowing across an international border
- Type (e): a confined aquifer, unconnected hydraulically with any surface body of water (except perhaps within the recharge zone in an unconfined portion of the aquifer) that traverses an international boundary or that is located completely in another state;
- Type (f): transboundary aquifers that are unrelated to any surface body of water and disconnected from the hydrologic cycle, thus devoid of any meaningful recharge.

In transboundary situations, depending on the location of the international border, sedimentary and alluvial aquifers may be classified in four different types, as follows (Fig. 1.18, Chilton 2007, Personal Communication):

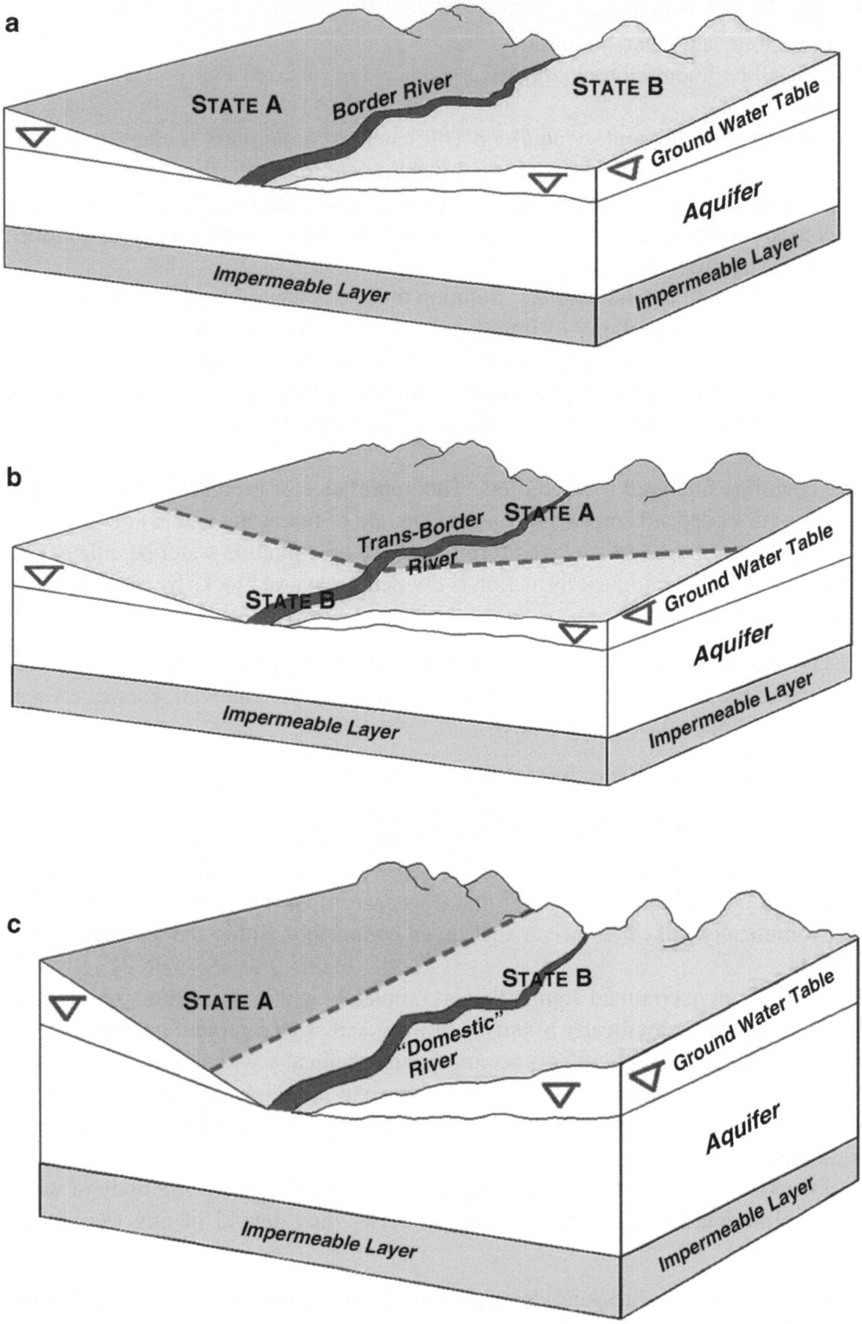

Fig. 1.17 Models of transboundary aquifers (Fried and Ganoulis 2016)

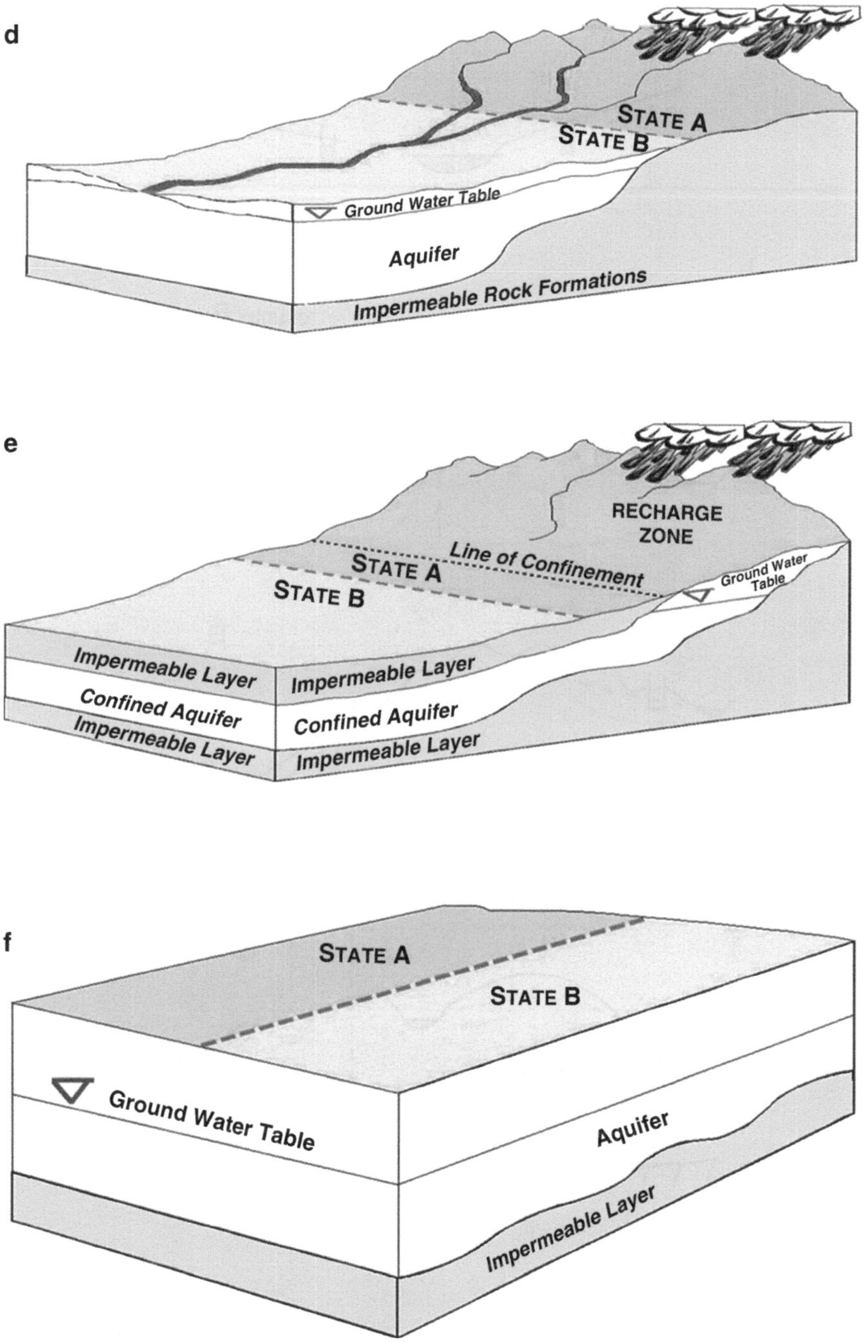

Fig. 1.17 (continued)

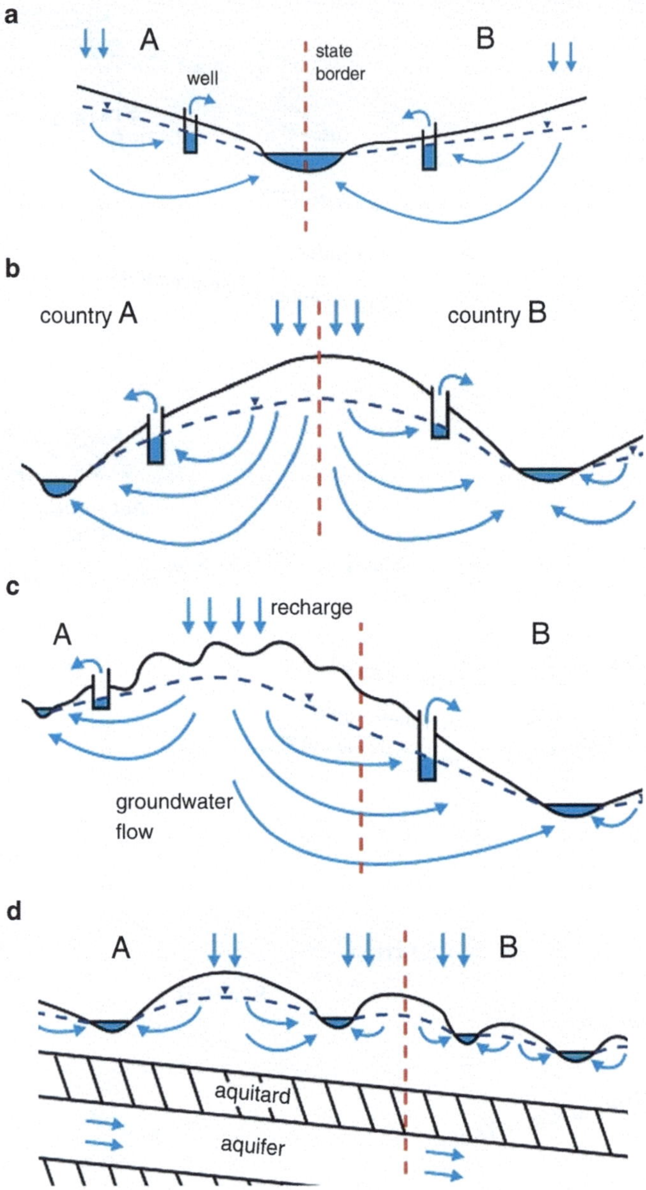

Fig. 1.18 Types of transboundary sedimentary aquifers (Chilton 2007)

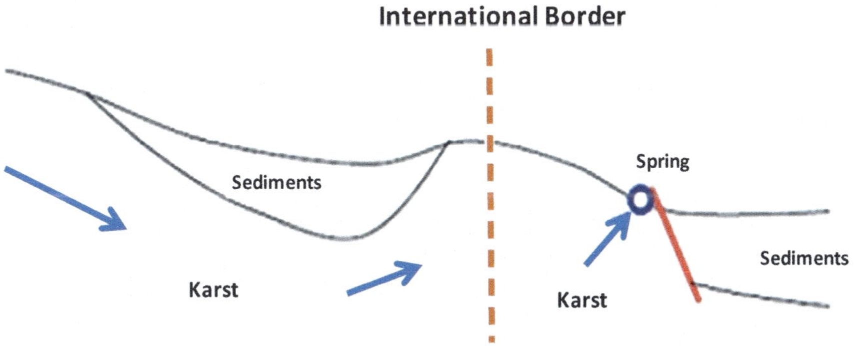

Fig. 1.19 Large deep karst aquifer recharged in one country and forming a spring in a neighbouring country

- Type (a): the state border follows a transboundary river or lake. Little transboundary groundwater flow occurs in the alluvial aquifer connected to the river;
- Type (b): the state border follows the basin and groundwater is divided. Very limited discharge occurs across the border;
- Type (c): the state border is separate from the basin and groundwater is divided. Recharge occurs in one country and discharge in the other;
- Type (d): Large deep aquifer recharged far from the border. Transboundary groundwater flow not connected to the surface may be important.

In the case of deep karst aquifers covered by sediments, recharge of groundwater may occur in one country and water can appear at the surface in the form of a spring in another country (Fig. 1.19). This occurs frequently in the Dinaric karst (Western Balkans), between Bosnia and Herzegovina (upstream) and Croatia (downstream, near the Adriatic coast).

References

Bodelle J, Margat J (1980) L' Eau Souterraine en France. Masson, Paris

Chilton J (2007) Personal communication

Fried J, Ganoulis J (eds) (2016) Transboundary groundwater resources: sustainable management and conflict resolution. Lambert Academic Publishing, Saarbrücken

Ganoulis J et al (eds) (2011) Transboundary water resources management: a multidisciplinary approach. Wiley-VCH, Weinheim

Gleick PH (1996) Water resources. In: Schneider SH (ed) Encyclopaedia of climate and weather, vol 2, Oxford University Press, New York, pp 817–823 and USGS (US Geological Survey). https://water.usgs.gov/edu/watercycle.html. Accessed 20 Dec 2017

IGRAC (2017) International groundwater resources assessment centre, transboundary aquifers of the world map 2015. https://apps.geodan.nl/igrac/ggis-viewer/viewer/tbamap/public/default. Accessed 20 Dec 2017

Musy A (2001) e-drologie. Ecole Polytechnique Fédérale, Lausanne. http://echo2.epfl.ch/VICAIRE/. Accessed 20 Dec 2017

Quante M (2007) Distribution and transport of water in the atmosphere. In: Lozán JL et al (eds) Global change: enough water for all? Wissenschaftliche Auswertungen und GEO-Verlag, Hamburg, pp 45–52

TWAP (2016) Transboundary waters assessment programme (GEF TWAP). http://www.geftwap.org/ and http://ihp-wins.unesco.org/layers/geonode:transboundary_river_basins and http://twap-rivers.org/. Accessed 20 Dec 2017

UN-Water (2018) Transboundary waters, facts and figures. http://www.unwater.org/water-facts/transboundarywaters/. Accessed 30 April 2018

UN WWDR3 (2009) Facts and figures, UNESCO Publishing, 7, Place de Fontenoy, Paris. http://www.unesco.org/new/en/natural-sciences/environment/water/wwap/facts-and-figures/. Accessed 20 Dec 2017

UNESCO glossary (2012) International glossary of hydrology, UNESCO, WMO No 385 at http://unesdoc.unesco.org/images/0022/002218/221862M.pdf. Accessed 20 Dec 2017

UNESCO/ISARM (2001) A framework document. UNESCO, Paris, Non-serial documents in hydrology

USGS (2017) US geological survey: groundwater discharge – the water cycle. https://water.usgs.gov/edu/watercyclegwdischarge.html. Accessed 20 Dec 2017

WHYMAP (2008) World-wide hydrogeological mapping and assessment programme, BGR, UNESCO. https://www.whymap.org/whymap/EN/Home/whymap_node.html. Accessed 20 Dec 2017

Wolf A et al (1999) International river basins of the world. Int J Water Resour Dev 15(4):387–427

Chapter 2
Transboundary Water Security

When the well is dry, we know the worth of water.
Benjamin Franklin, Founding father of the US

Abstract Water is a key factor for the socio-economic development of human societies as it is the main element for human and ecosystems survival, and it plays a major role in different economic activities, such as agricultural irrigation, energy, industrial production and environmental preservation. Various natural and man-made factors, such as regional and temporal water variability, physical or economic water scarcity, are sources of potential water conflicts between neighbouring countries sharing common surface and groundwater resources. In this chapter, starting with the study of water variability, both in space and time, the concept of *water security* in transboundary river catchments is analysed. This means that countries sharing the same surface and groundwater resources should ensure enough water in all circumstances, such as those of low flows and minimise the risk of flooding when water exceeds the carrying capacity of the flood ways. Water security should be ensured not only under seasonal and regional water variability but also under climate change, which generates hydrologic instabilities, i.e. the frequent occurrence of maximum and minimum extreme flows.

Keywords Transboundary water variability · Scarcity · Security · Climate change · Floods · Droughts

2.1 Water Variability

2.1.1 Spatial and Temporal Water Variability

Freshwater is unequally distributed around the globe and also shows a high temporal variability. Enormous arid zones on the planet are totally lacking water and are prone to droughts, while other regions are over-saturated and vulnerable to floods. This variability obviously affects transboundary areas, where countries share common

J. Ganoulis, J. Fried, *Transboundary Hydro-Governance*,
https://doi.org/10.1007/978-3-319-78625-4_2

surface and groundwater resources. For example, countries with temperate climate and sufficient national water resources could possibly face the reality of uneven water supply from neighbouring countries due to climate change.

Spatial Variability

Water is not systematically concentrated in areas where anthropogenic needs are important. For example, the Great Lakes in North America, which represent the largest proportion of the world's freshwater reserves (21% of the world's surface freshwater), supply only a very small number of the world's inhabitants. Arid and semi-arid regions representing 40% of the earth's surface lands receive only 2% of the world's runoff water (Fig. 2.1).

The regions that benefit from sufficient water are those where the mean annual precipitation exceeds 500 mm/year. In Fig. 2.2 these regions are shown on the map

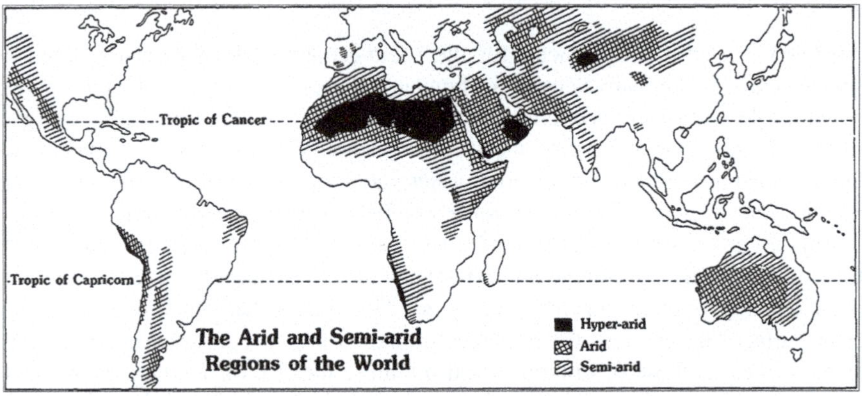

Fig. 2.1 Distribution of arid and semi-arid regions on Earth (UNESCO 1979)

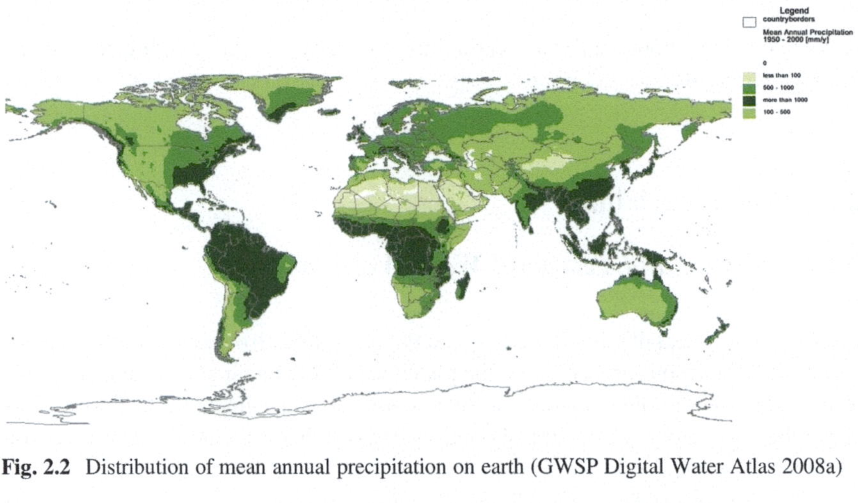

Fig. 2.2 Distribution of mean annual precipitation on earth (GWSP Digital Water Atlas 2008a)

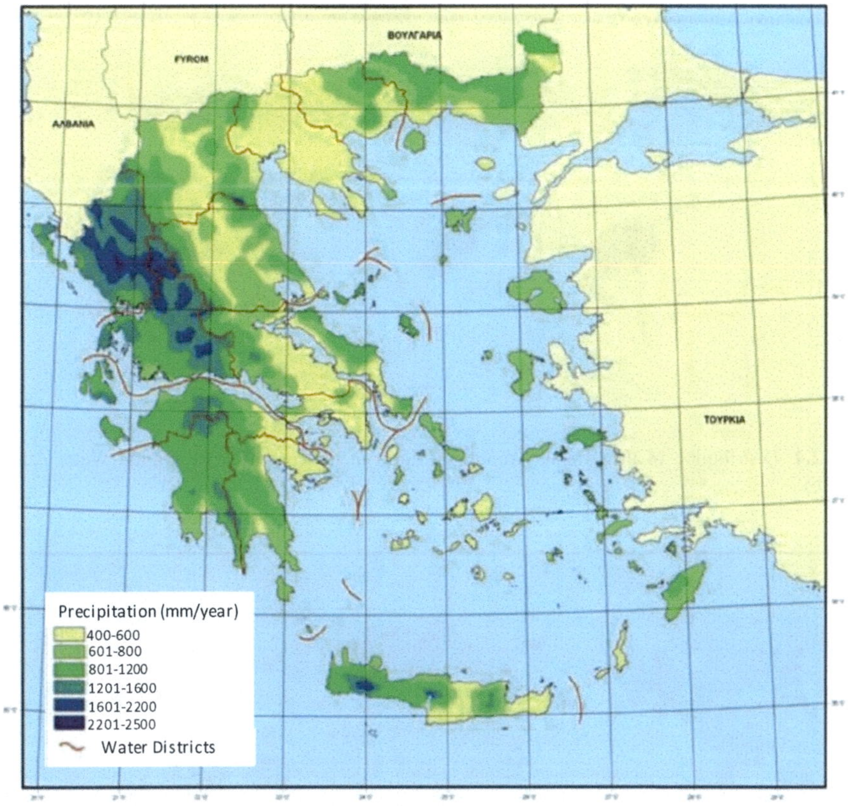

Fig. 2.3 Distribution of mean annual precipitation in Greece

with dark to very dark and are areas where the climate varies from temperate or Mediterranean to humid, tropical and rainforest. In these regions, rainfall's spatial variability is also important and is significant even at smaller scale, as shown in Fig. 2.3 for the mean annual precipitation in Greece.

For water resources management and socio-economic development, it is not the precipitation itself which is of importance, but rather the so-called *efficient precipitation*. This is the amount of water resulting when the *evapo-transpiration* is subtracted from the total precipitation. This amount represents the surface runoff plus the water infiltrating into the soil (*renewable groundwater*). The distribution of *surface runoff* and *renewable groundwater* in mm/year are shown in Figs. 2.4 and 2.5.

The complexity of runoff distribution across the globe is illustrated in Fig. 2.4. High runoff values are observed in regions of high precipitation, tropic climate and high moisture conditions. Runoff is lowest in the arid regions, such as the Saharan region of Africa, central Asia and Australia and regions of South America. Of special importance are the transitional zones between humid and arid zones, such as the Sahelian region of Africa.

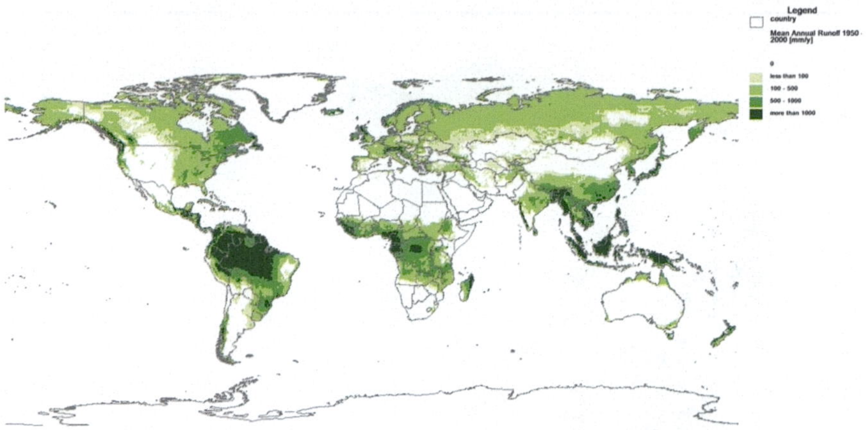

Fig. 2.4 Distribution of mean annual runoff on Earth in mm/year (GWSP Digital Water Atlas 2008b)

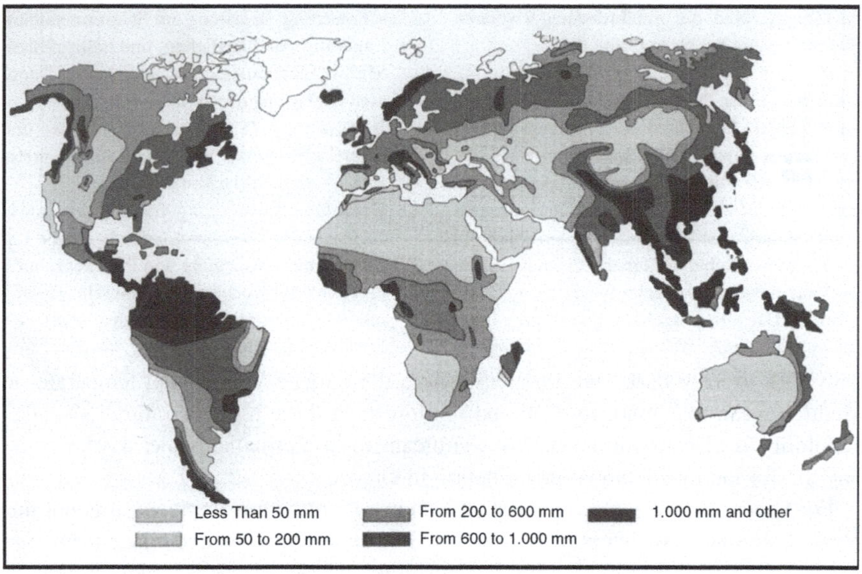

Fig. 2.5 Depth of mean annual groundwater discharge (WRI 1990)

This regional distribution of water resources across the globe is also affected by temporal variability that may be periodical, such as seasonal and the monsoon phenomena in India or non-periodical as this due to climate change. Areas affected by climate change may experience higher or lower precipitation and in the same time, atmospheric instabilities due to greenhouse gases may induce more frequent and intense extreme hydrological events, such as floods and droughts.

The distribution of renewable groundwater resources in mm/year is shown in Fig. 2.5. In this figure, the total flux through the aquifers over a long period of time is estimated by the groundwater discharge or the annual groundwater outflow. In most cases, groundwater is preferentially used for drinking and irrigation purposes because it has a much better quality than surface water, as it is protected from pollution by geological formations. Hence, groundwater is an important and precious freshwater resource, and its value is expected to increase further in the future.

As shown in Fig. 2.5, the mean annual renewable groundwater may be in some aquifers low (less than 50 mm/year), medium (50–200 mm/year) and high (>200 mm/year). Compared to the average annual recharge of aquifers, these renewable natural resources are very varied and unevenly distributed. For example, in the Mediterranean Basin with a total amount of about 155 km^3/year, 93% is located between its northern shores (South Europe and Turkey) and only 7% in the southern shores. These recharges vary from a few mm/year in the arid zone of the south up to more than 500 mm/year in the Dinaric Alps (Aureli et al. 2008).

The majority of these underground flows contribute to the regular flow rate of the water streams, therefore constituting a constant resource to surface water. A significant part ends up in near-shore submarine springs or in coastal karstic aquifers. According to technico-economic criteria these resources are unevenly accessible and exploitable, and also commensurate with the constraints applied for the conservation of surface flow. They are also sometimes relevant to the preservation of tributary aquatic ecosystems of the emerging underground water.

When comparing water demand with total available water resources, it can be seen that in many countries the opportunity to satisfy water needs has become a serious problem. The pressures resulting from groundwater exploitation are varied, often heavy and even excessive in numerous cases. Over-exploitation is frequent and increasing, notably in the coastal aquifers, where it has resulted in marine water intrusion. This process will be difficult to reverse, both in the north and the south. It has also caused streams to dry up and has weakened or destroyed aquatic ecosystems. Taking into account both this level of over-exploitation and the mining of 'fossil water', it can be said that the current extraction of groundwater in many areas is not sustainable.

These strong pressures on resources often provoke conflicts in usage, including conflicts between users of groundwater and surface water. Moreover, in numerous areas, the quality of groundwater is under threat from pollution, which is especially linked to urbanisation and the impacts of intensive agriculture. This situation also constitutes an area of conflict between users of groundwater and land owners. Satisfying the need for high quality drinking water represents a particularly complex problem. The rational use of groundwater can play an important role in solving this problem.

An important indicator for the sustainable use of the total renewable water resources, both surface and groundwater, is *water availability*, that is the total annual renewable amount of surface and groundwater which is available to a person. If, in a country, water availability is less than 1000 m^3/year/person we can say that this country experiences *water scarcity*. As shown in Fig. 2.6, for water availability

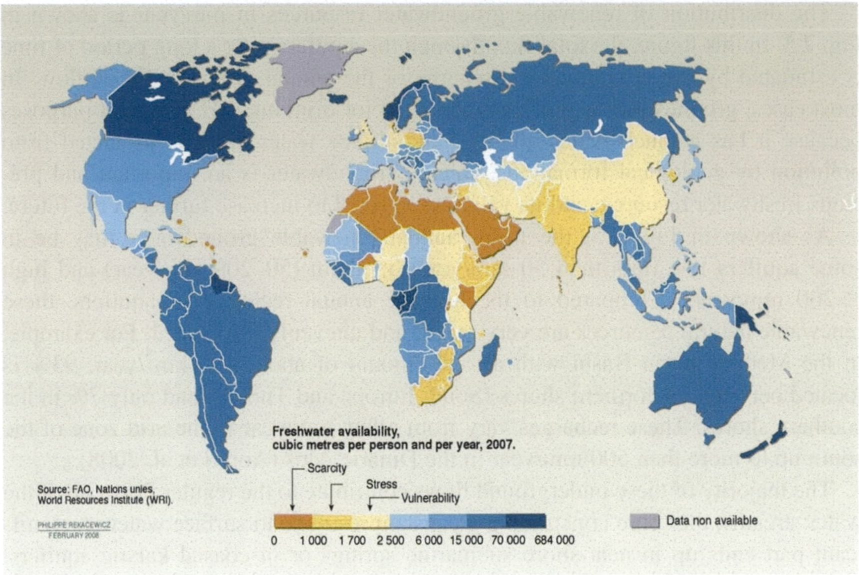

Fig. 2.6 Distribution of the mean freshwater availability (m^3/year/person) (WRI 1990)

between 1000 and 1700 m^3/year/person, the country faces water stress, and it is vulnerable for water availability between 2700 and 1700 m^3/year/person. Of course, the above figures are not fixed with great precision, they may vary in time and only offer an indication about the degree of importance of the water availability for human use.

Of particular importance for the management of national water resources is the ratio between the amount of transboundary water resources and the national or internal water resources, both surface and groundwater. This indicator is called *dependency ratio* and as shown in Fig. 2.7, it varies from 0 to 100%. The map shows how much a country depends on neighbouring countries for its total water resources, or the contribution of transboundary water to the total country's renewable water resources. As we can see, most of the countries share water resources with their neighbours up to certain degree. The dependency ratio is approaching extreme values, such as 100% in the case of Egypt (Nile and the Nubian Sandstone Aquifer) and 0% for Australia, Japan, Indonesia and other island states (Fig. 2.7).

Management of shared water resources should be based not only on reliable scientific knowledge but at the same time on good cooperation between neighbouring countries. This applies also to transboundary or shared aquifer systems that are important sources of fresh water in many regions of the world, such as in the MENA (Middle East and North Africa) region, as well as in South America (the Guarani Aquifer, shared by Argentina, Brazil, Paraguay and Uruguay).

Developing cooperative databases for sharing information from different sources and mainly from regional monitoring networks is a prerequisite for formulating and

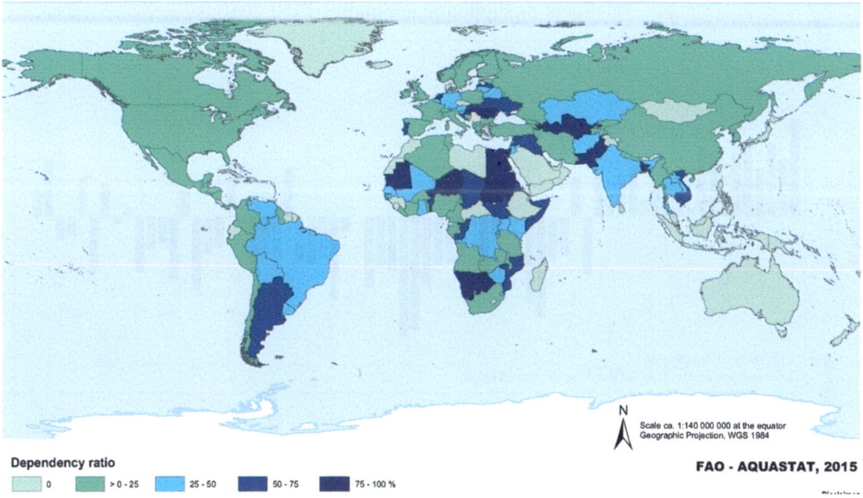

Fig. 2.7 Distribution of the *dependency ratio* between shared and national water resources (%)

implementing common strategies and management policies for shared water resources. Collecting data at rates that can be sustained over long periods of time, such as water levels, water extraction and groundwater quality in an aquifer is the foundation of groundwater management. In parallel to reliable data collection, the organisation of databases for analysing information and data on water resources, in terms of quantity and quality, are vital to efforts directed towards planning in order to meet present and future water demands.

Temporal Variability

Water resources are also unevenly distributed in time. Many parts of the world experience pluri-annual periods of low precipitation or in the opposite periods of water scarcity and droughts. The story of the 7 years floods of the Nile River, in ancient Egypt, followed by 7 years of low waters is well known. Joseph interpreted the Pharaoh's two dreams as 7 years of good agricultural production, followed by 7 years of famine and this provides a striking example of long-range planning based on a climate forecast. Based on this forecast, Joseph proposed to Pharaoh a plan for running the agricultural economy of Egypt. In fact, ancient Egypt's rural economy was entirely dependent on the Nile River's seasonal flooding: when the highest water levels covered the arable areas of the river valley, crops were sufficient to feed the population. In the opposite, during the successive years of low flow of the Nile, lack of flooding resulted in low agricultural production, while hunger weakened the economy and the state.

Similar long term variability of annual precipitation is also observed over the Sahel of Africa (See: University of Washington, Joint Institute for the Study of the Atmosphere and Ocean: http://research.jisao.washington.edu/data_sets/sahel/. Accessed 20 December 2017). The Sahel forms a border between the Sahara desert to the north, at about 18°N, and the savannah and tropical forest to the south, at about

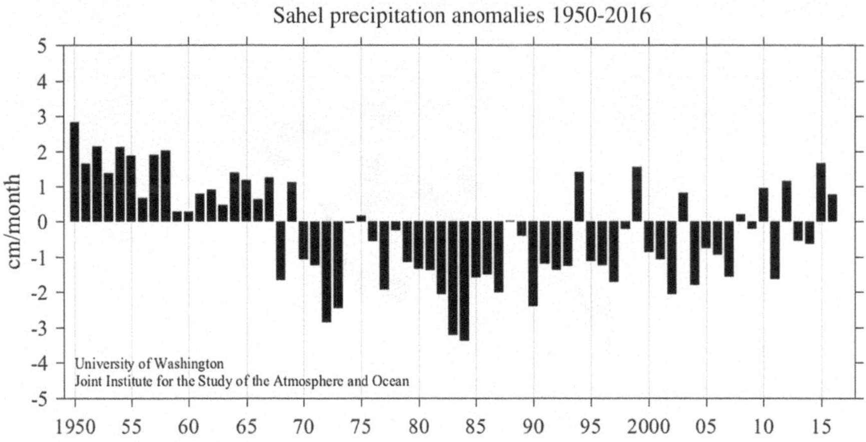

Fig. 2.8 Time series of monthly precipitation in the Sahel (1950–2015)

15°N. During the summer, when the ITCZ (Inter Tropical Convergence Zone, a belt of low pressure that circles the Earth near the Equator) moves northward, the region is affected by the south-westerly monsoon that brings rain. The ITCZ then shifts south, and the predominant winds become northerly (from the desert) resulting in aridity. This region became significantly more arid during the second half of the twentieth century, especially after the 1960s. The situation reached its peak during the 1970s and 1980s and in recent years, the trend seems to have reversed, as shown in Fig. 2.8.

Seasonal Variability

Precipitation and river flow also show seasonal variability due to seasonal weather variation. Well known is the monsoon phenomenon, which refers to a seasonal shift of winds due to land-ocean temperature contrasts. Monsoon phenomena occur on many parts of the world, such as the south-western U.S., but the best known of these is the Indian monsoon. In summer, low pressure develops over land, resulting in an inflow of moist air which promotes precipitation; in winter, there is an outflow from the colder land, promoting dry conditions.

Depending on the location of a river catchment, the climate of the region and the altitude of the springs alimenting the river, different seasonal regimes can be distinguished.

The *nival regime* is similar to the *glacial*, but attenuated and the maximum flow takes place around June–July. It can be mountain, as in the case of Mesta/Nestos (Fig. 2.9, Skoulikaris 2009) or plain nival regime. The characteristics of the plain nival are: (a) floods in May–June and (b) great variability daily, yearly and inter-annually.

The *pluvial/oceanic regime* is typical of rivers at low to moderate altitude (500–1000 m) and is characterized by: (a) high water in winter and spring, (b) low discharge in summer and (c) great inter-annual variability.

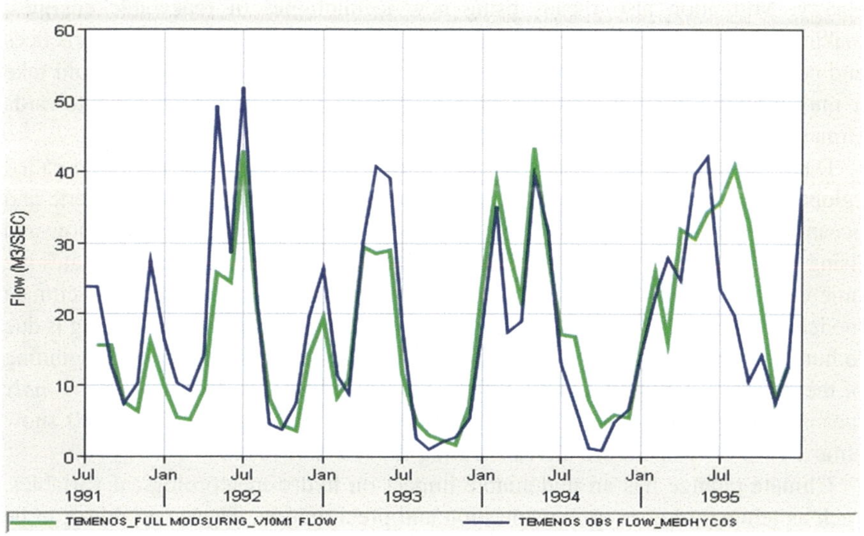

Fig. 2.9 Measured and simulated flow rate variation of Mesta/Nestos River (mountain nival regime)

The *tropical pluvial* regime is characterized by: (a) very low discharge in the cold season and abundant rainfall in the warm season, (b) minimum can reach very low values, (c) great variability of discharge during the year and (d) relatively regular from 1 year to another.

2.1.2 Climate Change: Mitigation and Adaptation

Climate change refers to the modification of the actual climate humans experience on earth, due to the extensive emission of greenhouse gases, such as CO_2 and CH_4. In order to face this major challenge that humanity faces, there are two different but complementary approaches: *mitigation* and *adaptation*.

Mitigation

Mitigation means the possible reduction of greenhouse gases in order to decelerate and possibly stop their negative impact on the climate. This is a long-term common effort of all countries that have ratified the UN Framework Convention on Climate Change (UNFCCC) in 1992 at the Earth Summit in Rio de Janeiro and agreed in 2015 during the COP21 (twenty-first Conference of the Parties) in Paris to adopt the necessary measures to address climate change and its impacts on our society. These measures consist mainly of cutting down the greenhouse gases by replacing the energy production from fossil sources, such as coal and oil, by renewable sources of energy. Mitigation also means using new technologies of renewable energies, making older equipment more energy efficient, and changing management practices and consumer behaviour. Countries willing to combat climate change should take a multifaceted approach towards climate change mitigation and move towards climate-resilient and low emissions strategies.

Data from around the world show that during the last few centuries the so-called "global warming" has occurred, i.e. there is a constant increase in atmospheric and oceanic temperatures (Figs. 2.10 and 2.11), widespread melting of ice and snow and rising of global sea level. Despite the high degree of uncertainty that dominates the time variability of the Earth's climate, recent studies on climate change and climate modelling indicate that there is at least a 90% probability that global warming is due to human activities and more specifically to gaseous emissions since the beginning of the industrial revolution in 1750 (IPCC 2001). If the term *temperature anomaly* means the departure from a reference value, mean global data in Fig. 2.10 show almost 1 °C of temperature increase during the last 136 years.

Climate change has an undeniable impact on hydro-meteorological variables, such as temperature, evapotranspiration and precipitation. These variables can be calculated by using different *climate models* that simulate global atmospheric circulation by use of mathematical equations expressing the mass, momentum and energy balance governing the atmosphere and the circulation behaviour of the oceans. These models are derived from the latest advances in fluid dynamics and thermodynamics taking into account chemical interactions. However, these non-linear partial differential equations can only be solved using approximate numerical solutions. These solutions are very sensitive to the correct setting of initial boundary equations. Atmospheric temperature fluctuations are induced by variabilities and instabilities of different origin, such as air turbulence and density distribution. The following three causes are also responsible for these fluctuations:

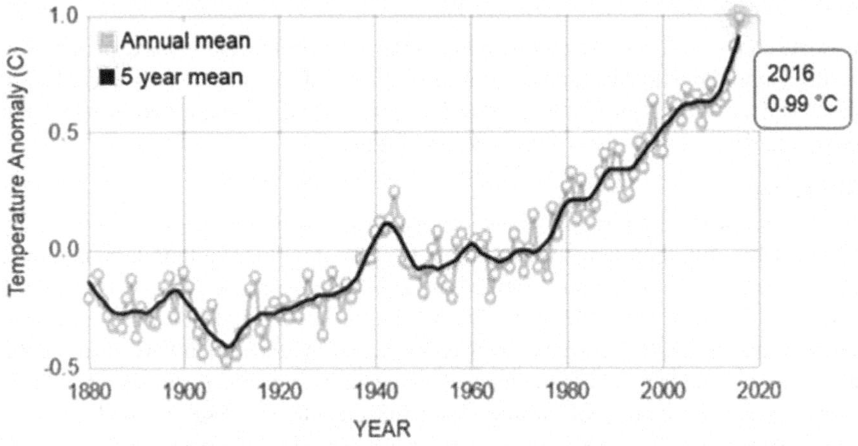

Fig. 2.10 Global temperature anomaly variation (1870–2016, 136 years)

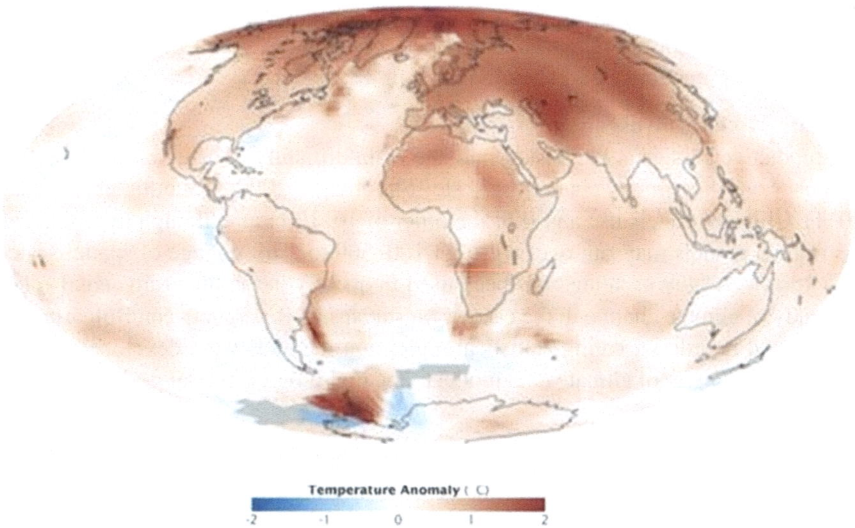

Fig. 2.11 Mean temperature anomaly 2000–2009 (NASA earth observatory)

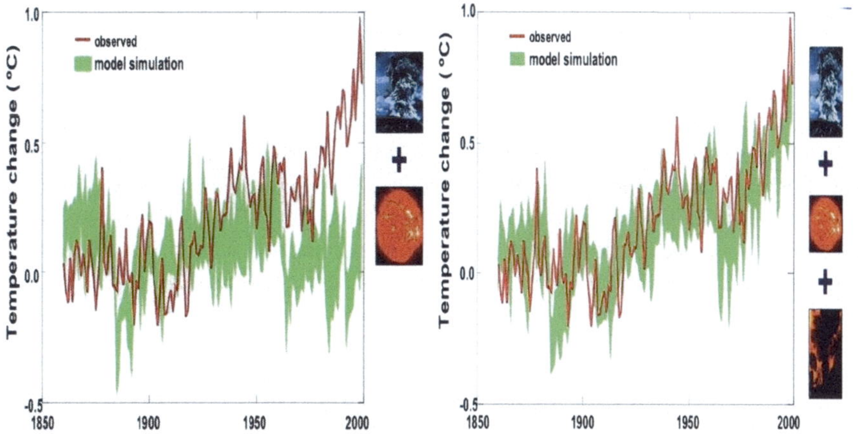

Fig. 2.12 Observed and simulated temperature variation (**a**) without and (**b**) by taking into account human gaseous emissions (Hadley Centre 2005)

(a) Temporal changes in solar energy,
(b) Volcanic emissions of aerosols, and
(c) Gas emissions from anthropogenic activities, mainly from fossil-fuel burning.

As shown in Fig. 2.12, climate models are able to predict temperature increase after 1950 only if the man-induced gas emissions are taken into account. In fact, in

Fig. 2.12(a) changes in solar energy and volcanic emissions are only taken into account and the model can reproduce temperature increase after 1950 only if human emissions of greenhouse gases are included in the model.

Climate Model Downscaling

For predictive studies at the basin scale, the main difficulty in using climate models at local level lies in the change of scale (Skoulikaris 2009; Ganoulis et al. (eds) 2011). Climate models use a grid of the order of 10–100 km, although hydrological distributed models may run over a grid of 100–500 m. This change of spatial scale, also known as "downscaling" has been used frequently in the literature and also for the study of climate change in the case of Mesta/Nestos River, between Bulgaria and Greece. Ad hoc downscaling statistical regression methods were developed in this case for the transfer of climate data model to the hydrological model grid (Fig. 2.13).

Emission Scenarios

Since its creation, the IPCC (International Panel on Climate Change) has regularly prepared assessment reports on available scientific information on climate change and has proposed possible response strategies. The First IPCC Assessment Report (FAR) was presented in 1990 and was devoted to creating an inventory of climate change. The Second Assessment Report (SAR) was presented in 1995 and was coupled with the initial Special Report on future Emissions Scenarios (SRES). The future emission scenarios, also known as IS92, were subsequently used in climate simulation models in order to quantitatively assess their impacts. A follow-up Third Assessment Report (TAR) was presented in 2001 and the updated version of the SRES scenarios of the TAR was definitively adopted in the Fourth Assessment Report (AR4), which was presented in 2007. The global CO2 emissions for the six SRES scenarios, A1B, A2, B1 and B2, A1FI and A1T are presented in Fig. 2.14. The IS92a scenario derived from the Second Assessment Report (SAR) is also presented. Obviously, the most optimistic scenarios are A1T and B1, where clean energy technologies are developed. At the other extreme, the scenarios A1FI and A2 demonstrate a huge increase in CO2 emissions, since economic development is not coupled with environmentally friendly policies and technologies.

Different atmospheric emission scenarios have been used for studying the impacts of climate change on river hydrology. The Mesta/Nestos River Basin has been selected as a pilot basin for the following reasons: (1) it combines internationally shared surface and ground waters, (2) there are serious issues of hydrological and environmental concern in the delta region and (3) competing water uses have been increasing from the time when Greece built a series of hydro-electrical plants in the downstream part and Bulgaria entered a market economy and embarked on a struggle for a rapid economic growth.

Figure 2.15 presents a comparison between the flow computed upstream of the Thissavros dam on the Nestos River for the IPCC A1B climate scenario and a reference flow series prolonging in time the climate conditions of the past 25 years. It is worth noting that this reference series contains similar dry spells to the ones experienced in the area in the early 1990s. This climate stability hypothesis is also sometimes referred to as "business as usual" or "baseline scenario" in the EU Water

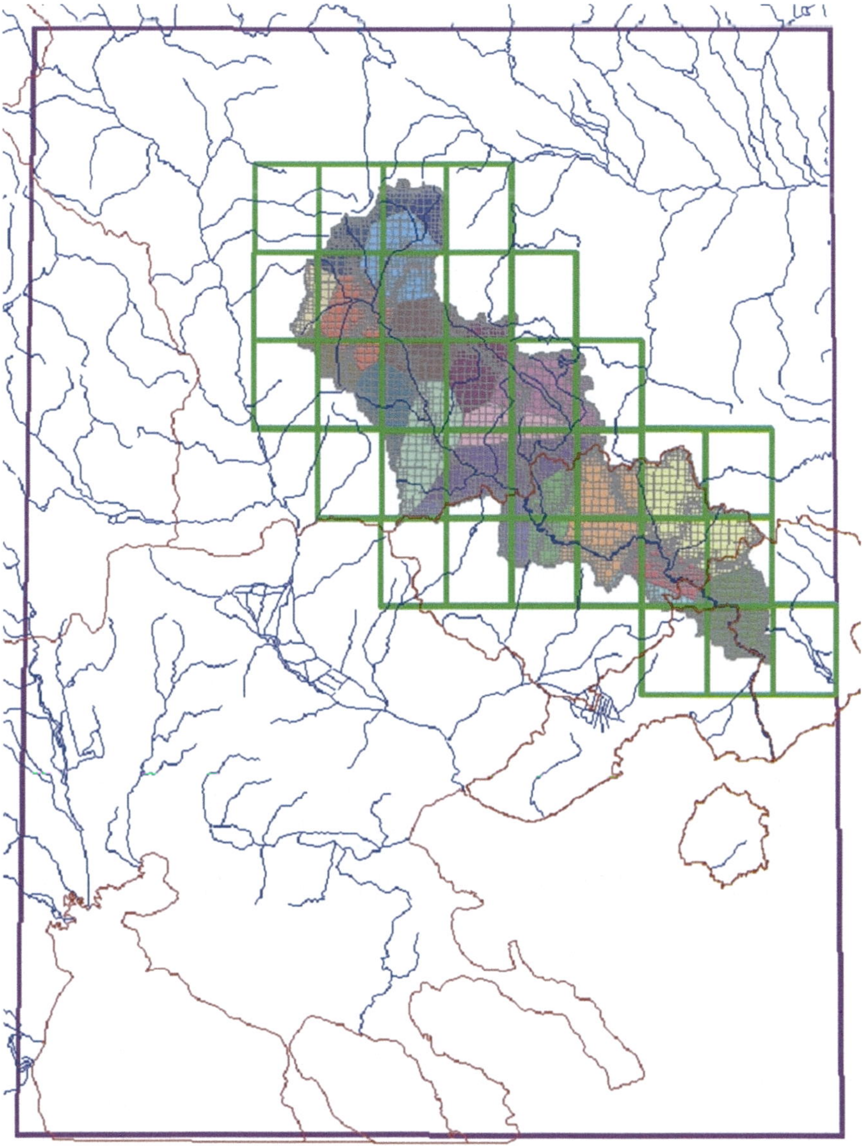

Fig. 2.13 Downscaling from the climate model (20 × 20 km) to the hydrological grid

Framework Directive (WFD) nomenclature. Compared to climate conditions in the recent past, the A1B scenario clearly predicts a flow decrease.

Impacts of climate change on the water cycle could mean that some regions will become wetter and others dryer and they possibly turn into arid and semi-arid regions. Modifications in the water cycle will impact the health of fragile ecosystems

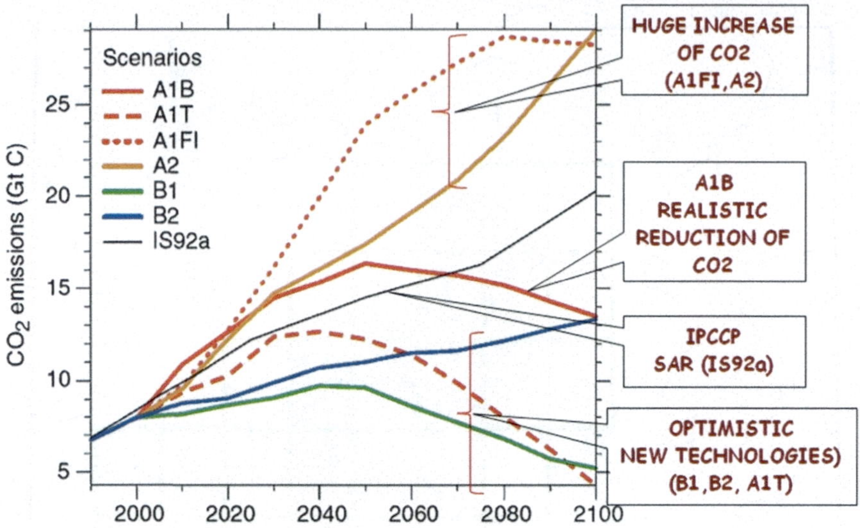

Fig. 2.14 Different scenarios of future CO_2 emissions

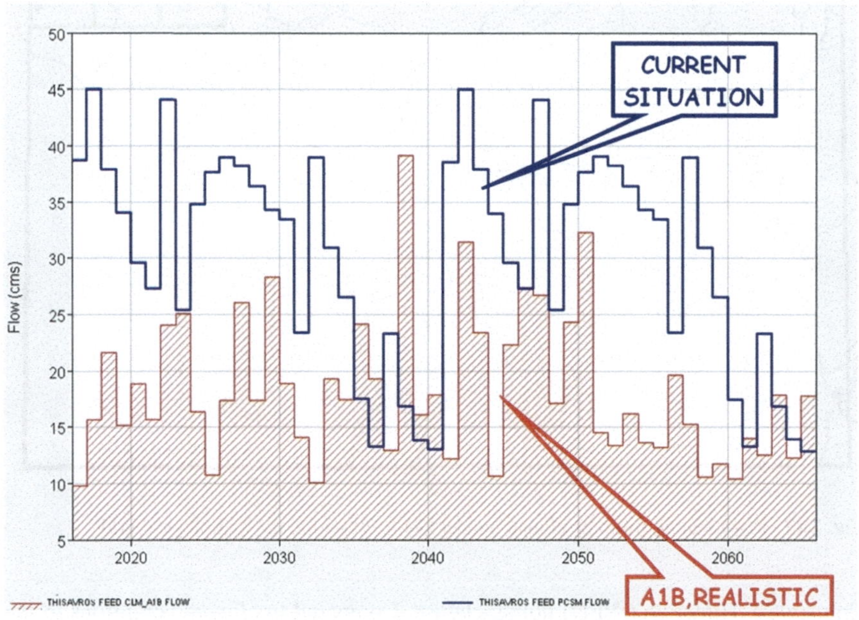

Fig. 2.15 Comparison between yearly river flow simulated for the IPCC A1B case and a prolongation of past climate conditions (baseline scenario, current situation)

in these regions, and consequently endanger the lives of people who depend on the natural resources that these ecosystems provide. Today already, more than 40% of the world's land resources are in dry lands (i.e. sub-humid, semi-arid and arid regions) that are threatened by land degradation and desertification. According to the United Nations Convention to Combat Desertification, this is seen to have a direct impact on 250 million people and an indirect impact on more than a billion. Quite inevitably, changing water cycles as a result of climate change are expected to affect survival of plant and animal species, land use, food production, and thus the so far existing balance between man and its surrounding environment.

Adaptation

Estimation of the residence time of greenhouse gases into the atmosphere and especially that of CO_2 varies from few decades up to 100 years. In any case, as these gases remain in the atmosphere for a long time after their emission it would take too long before the mitigation measures show effects. Consequently, *adaptation* to climate change is necessary and urgent since the climate is already changing in some respects. Even if global mitigation efforts prove successful, further climate change throughout this century and beyond is almost certain. In addition, it is more cost-effective to start preparing for adaptation now than delaying it for confirmation of more negative impacts. As shown in Fig. 2.16, the mean precipitation pattern over the Mediterranean Region shows an important reduction up to 50% by the end of this century.

Adaptation to climate change represents an important challenge for all countries, especially for those with economies in transition, but few of them have developed adaptation strategies so far. Especially knowledge and experience of adaptation in a transboundary context are lacking. In transboundary areas adaptation is much more complicated. For example, in the UNECE region climate change will affect more

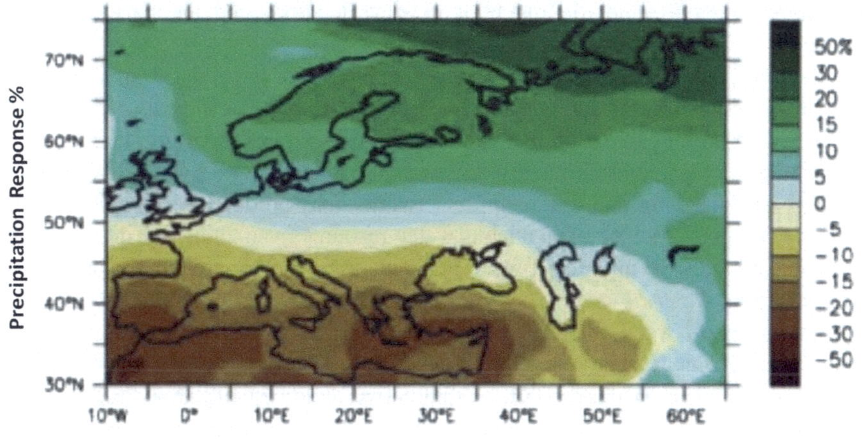

Fig. 2.16 Percentile changes (averaged over 21 models) in annual mean precipitation between the periods 1980–1999 (observed precipitation) vis-à-vis the period 2080–2099 (expected precipitation)

than 150 transboundary rivers, 50 major transboundary lakes and more than 170 transboundary groundwater systems, which means that riparian countries should produce cooperative solutions.

For this reason, the Meeting of the Parties to the UNECE Water Convention, at its fourth session in Bonn, Germany, 2006, decided to assist Governments in developing adaptation strategies at different government levels. Following this decision, a Guidance Document was prepared by the Task Force on Water and Climate under the Water Convention, in close cooperation with the Task Force on Extreme Weather Events, under the Convention's Protocol on Water and Health (UNECE 2006) (Fig. 2.16).

2.2 Transboundary Water Scarcity

2.2.1 Types of Water Scarcity

Water scarcity already affects every continent. Around 1.2 billion people, or almost one-fifth of the world's population, live in areas of physical scarcity, and 500 million people are approaching this situation. Another 1.6 billion people, or almost one quarter of the world's population face economic water shortage (where countries lack the necessary infrastructure to take water from rivers and aquifers).

Water scarcity is among the main problems to be faced by many societies and the World in the twenty-first century. Water use has been growing at more than twice the rate of population increase in the last century, and, although there is no global water scarcity as such, an increasing number of regions are chronically short of water. Water scarcity is both a natural and a human-made phenomenon. There is enough freshwater on the planet for six billion people but, as we have seen in Sect. 2.1, it is distributed unevenly and too much of it is wasted, polluted and unsustainably managed.

Water scarcity occurs due to different reasons. Arid or semiarid areas in low latitude are vulnerable to water scarcity because of a lack of sufficient precipitation together with high evaporation that can lead to *natural water scarcity.* In countries with low economic possibilities low investment capabilities in water related infrastructure, such as dams and water supply networks, may produce an *economic water scarcity.* Mismanagement of available water resources can induce a demand for water that exceeds the amount of water available in one given area: in that case, we may have a *managerial water scarcity.*

In some cases, water scarcity is declared when access to safe potable water is lacking, despite the total amount of water available. Therefore, a state water scarcity may be declared upon observation of qualitative and/or quantitative attributes of one given region's water offer. Nevertheless, scarcity often has its roots in water shortages, and it is in the arid and semiarid regions affected by droughts and wide climate variability, combined with population growth and economic development, that evidence of water scarcity is most present and ensuing consequences most acute.

Some experts have attempted to quantify water scarcity. According to certain studies, at levels between 1700 and 1000 m^3/year/person, periodic or limited water shortages can be expected. When water supplies drop below 1000 m^3/year/person, the country is said to be facing water scarcity. This method has however been criticised by some experts who deplore the fact that a variety of factors which also have an influence on water quantity and quality, and thus lead the "water scarcity range" to either being narrower or wider, are generally overlooked.

Water scarcity exists in several types, two of the most common being economic scarcity and physical scarcity. In the developing world, finding a reliable source of water is usually time consuming and can be very costly. This is known as economic water scarcity. Water can be found, it simply requires resources (wells, pumps, and pipes capable of channelling the water etc.) for people to start using it. Reasons for economic scarcity can be financial, political, with latent ethnic conflicts sometimes that hinder the effective construction or maintenance of water facilities. Economic scarcity is pointed out by some as being due to a lack of good governance as acted by politicians and decision-makers. Much of sub-Saharan Africa suffers under the effects of economic water scarcity, with people having to fetch water from rivers for domestic and agricultural uses being quite representative of their condition.

In some areas, the lack of water is a more profound problem: simply there is not enough for ecosystems to function effectively. That is known as physical water scarcity and it is a characteristic of naturally arid regions. Environmental degradation and declining groundwater quantities are usually the recurring symptoms observed in a state of physical scarcity.

Although physical water scarcity is often associated with dry parts of the world or arid climates, there is an increasing number of regions today where physical scarcity is a man-made condition. For instance, the Colorado River basin in the United States is an excellent example of a seemingly abundant source of water which has nonetheless been overused and over-managed, leading to a very serious case of physical water scarcity downstream.

2.2.2 Living with Droughts

Evidence and climatic projections suggest that water resources will be seriously affected by climate change in many areas prone to water scarcity and droughts. Given the inter-dependency between riparian states over shared water resources, only good cooperation between them may succeed in reducing drought vulnerability and impacts in transboundary river basins This means that gaps and particularities between the different governance structures are properly addressed and taken into account. There is a wide range of agreements between state governments on transboundary river basins, but most of them fail to achieve effective shared water governance (Molle et al. 2007). This is due to the fact that inter-state relations are embedded in complex political and economic frameworks, strongly centralised and bureaucratic, with poor stakeholder participation, and focused on retaining control of resources through their massive infrastructural exploitation (Merrey 2009).

Droughts may increase the potential conflict between countries sharing transboundary river basins as they are recurrent climatic hazards that cause major environmental, economic and social impacts. For example, the shared river basins of the Iberian Peninsula, and, the Sub-Danuvian river catchments in the Balkans (South Eastern Europe), are of major significance for Mediterranean Europe, as examples of basins marked by water scarcity and drought risk (Do 2012). Interstate political agreements that regulate water issues between riparian countries are applied with many difficulties under drought conditions. This is because usually riparian countries have different planning and management policies to tackle droughts, based on distinctive administrative and socio-economic structures. The level of cooperation and compliance with the basic rules of effective transboundary water governance can define potential constraints for shared and effective transboundary drought planning and management.

Most of the time, reality shows how riparian countries fail to identify common benefits and goals, based on a shared process of public participation, which still does not exist. The already developed experience shows that, before aiming at a shared or joint planning process, riparian countries need to incorporate adaptive and precautionary management procedures into their national planning systems and cooperation structures.

Most of the research on transboundary water governance has focused on formal legal instruments, conflicts and conflict resolution mechanisms at the national scale (Zeitoun and Warner 2006). However, little emphasis has been put on strategies and methodologies addressing effective shared governance, namely in critical situations such as droughts or other water-related risks and disasters (Videira 2007).

Droughts have singular features (unpredictability, slow and progressive onset, wide and blurred distribution both in time and space, non-structural and diffuse impacts) which favour a reactive rather than proactive behaviour.

Main issues to consider for an effective transboundary water governance of droughts are:

- setting priorities among economic activities (urban supply, livestock, permanent crops, ecological functions);
- bilateral compliance with European and international laws and regulations;
- need to inform the other party and conduct a transboundary impact assessment (based on the Espoo Convention principles) for any important water abstraction (above 40 hm^3/year);
- that water transfers to other river basins are restricted;
- the set-up of permanent information exchange circuits; and
- the promotion of a sustainable management of water (Serra 2000), since any significant increase in water consumption results in an increasing risk of non-compliance with the flow regime defined. Drought assessment indicators
- Drought management plans
- Prediction and early warning systems

- Impact monitoring and assessment
- Adaptation to climate change

2.3 Transboundary Floods

2.3.1 Flood Hazard and Risk Mapping

Extreme floods are essentially natural hazards that occur infrequently. In most cases, excessive precipitation is the main cause of catastrophic floods. However, anthropogenic factors, such as human occupation of floodways, extensive urbanisation and structural measures to mitigate floods (e.g. flood levees and walls, cutting of river meanders and river training) have modified the natural characteristics of extreme floods (Gardiner et al. 1995; Rossi et al. 1994). Several catastrophic floods in the past both in Europe and the USA (Rhine River flood in 1995; Elbe River in 2002; Mississippi River in 1993) have shown that human activity and traditional river engineering works may result in an increase in the frequency of extreme floods and the water stage with serious negative economic consequences, such as loss of or damage to property as well as danger to or loss of human life.

In transboundary river catchments depending on the position of the political boundary, floods may have different characteristics and different measures for prevention, mitigation and alleviation. In the case of upstream-downstream political division of the catchment area, floods are generated by the upstream country in its territory and exported into the downstream. This case is more difficult to resolve as an upstream country is in a more secure position during a flood. When the political border is located in the river's bed, floods generate symmetrical consequences and both countries are motivated for taking prevention measures.

In the Mediterranean area, flooding conditions in transboundary catchments are unique, given the influence of a semi-arid climate in many places, karst geological characteristics and the socio-economic environment. In this case, the main characteristics of floods may be described as follows:

1. The presence of heavy rainfall in autumn and winter may produce flash floods in catchments and streams, which remain dry throughout much of the year. These flash floods are of short duration (from a few minutes to a few hours) and have high flood peaks (many hundreds of m^3/s).
2. During flash floods, soil erosion and sediment transport are important and may lead to the failure of flood-defensive engineering structures (reservoirs, spillways, gates).
3. In karst areas, which make up more than half of the Mediterranean drainage basin, flash floods are more acute and much more violent (Davy 1989; Ganoulis 1995, 2009; Ganoulis and Vafiadis 1995). Excessive flooding occurs in these areas after the karstic cavities are filled by a huge amount of rainfall water.

4. Heavy concentrations of population in urban and residential areas around the centres of historic cities have, in many cases, resulted in the occupation of the beds and floodways of ephemeral streams. This phenomenon has been recorded mainly near the coastal areas, where tourist activity has dramatically increased in recent years. As the existing infrastructure in sewer systems is inadequate and its completion is very expensive, great volumes of storm water cannot be evacuated after heavy rains. Consequently, the lower areas of cities become flooded and serious damage to public and private property occurs (Ganoulis 1994).

In view of the limited economic means of local authorities, the implementation of traditional engineering measures to prevent floods, such as the building of dams and drainage tunnels, is very expensive. In populated areas, extension of the existing storm sewer system is not easy, due to the high cost of replacing the existing sewers and the impact of engineering works on urban activities such as trade, tourism and traffic. A risk-based design of alternative measures may be appropriate to reduce costs and to improve the reliability of the design (Ganoulis 2009).

Floodplain management and flood control involve *alternative measures* (*structural*: levees, dikes, retention basins, channel modifications, or *non-structural*: flood warning, land uses), *different natural conditions* (type of climate, socio-economic environment) and *various preferences* (economic, environmental, aesthetics, etc.). For the management of risks related to floods, various hydrological, economic and environmental uncertainties should be assessed and quantified. The US Army Corps Engineers' flood reduction structures use traditional principles of risk-based design, in that they only consider hydrological risk, which maximizes the net economic benefits from the project under various uncertainties.

In this context, the general principles of engineering risk analysis are used to develop a multi-objective risk-based approach to floodplain management. The various steps taken in a comprehensive application of engineering risk analysis to flood control are: (1) identification of hazards, (2) risk quantification, (3) consequences of risk and (4) risk management.

This multi-objective risk-based floodplain management approach is illustrated in the case of the Giofyros Basin near the city of Heraklion, in the island of Crete, Greece (Ganoulis 2003).

2.3.2 Flood Management and Early Warning Systems

Transboundary Flood Management

Many measures have been devised to help communities adjust to flood hazards and reduce the negative impacts of flooding. These include structural and non-structural, medium and long-term measures. Of the non-structural measures, complementary to all other forms of intervention, flood forecasting and early warning systems have proved again and again to be an effective and efficient tool for minimizing the negative impacts of floods, especially if we take into account that they are life saving.

Transboundary flood risk management in general, and specifically forecasting/early warning, has both a technical and a political aspect. In some countries, technical cooperation is ahead of institutional and political cooperation, i.e. it is not the technical capacity that is missing for common/integrated flood forecasting and early warning systems, but rather its transboundary institutionalization. In other countries, key problems are related to financing (often expensive systems) and type of processes (very complicated referring to flash floods).

Flood Risk Management requires adopting a river basin approach to planning through multidisciplinary inputs in order to reduce flood vulnerability and risks and preserve ecosystems. Flood risk management planning focuses on the reduction of potential adverse consequences of flooding for human health, the environment, cultural heritage and economic activity, on non-structural initiatives and on the reduction of the likelihood of flooding. To achieve this goal, flood risk management plans need to be developed to identify actions and measures to prevent and minimize the impacts of flooding.

Comprehensive flood risk management is crucial to reduce flood risks and it consists of key components that include (ISDR 2006):

1. *Prevention*: Preventative flood risk management towards preparedness, including spatial planning, the setting of flood defense measures and alarm systems, awareness raising campaigns among the population, etc.,
2. *Coping*: Flood management during events, implementing, forecasting frameworks and early warning (as described in the previous chapter), flood measures and evacuation plans; and
3. *Recovery*: Post flood event management, which includes aid, support and cleaning activities as well as the implementation of an appropriate assessment process to identify eventual shortcomings in existing flood management activities and plan improvement.

Early Warning Systems (from ISDR 2006)

Early warning systems focus on allowing individuals and communities threatened by hazards to react effectively (in sufficient time and in an appropriate manner), in order to reduce the impacts and damages of the hazard. To be effective and comprehensive, early warning systems should be composed of four inter-related elements: risk knowledge, monitoring and warning service, dissemination and communication, and response capability (see Fig. 2.17). All of the elements should be strongly interconnected and sustained by effective governance and institutional arrangements, including good communication strategies.

Risk Knowledge

This element aims at increasing knowledge about the risks individuals and communities face. Risk is a function of three factors: the magnitude of the hazard, the degree of exposure to the hazard and overall socio-economic and environmental vulnerability. Risk assessments should be carried out in advance to identify early

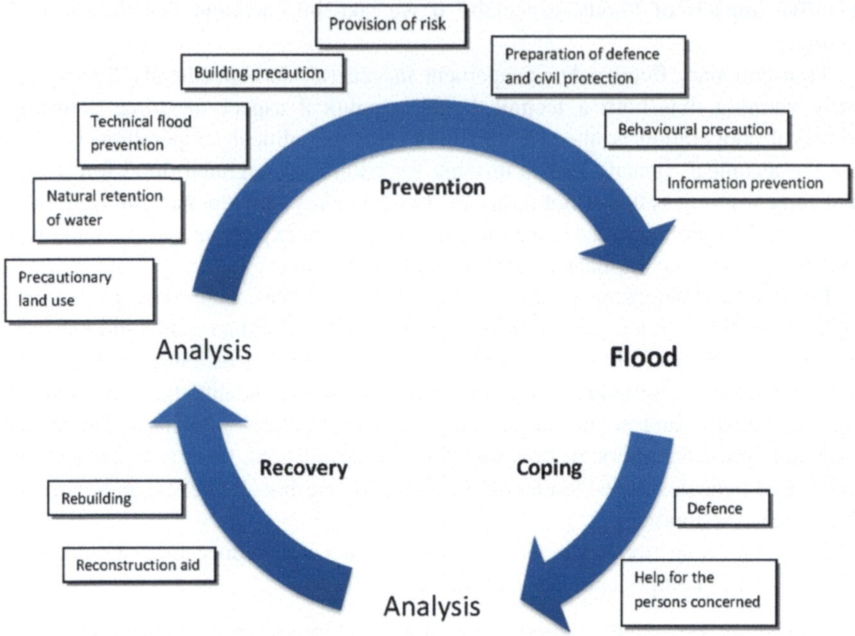

Fig. 2.17 Steps for flood risk management (from ISDR 2006)

warning system needs and prepare for response and disaster prevention activities. Risk assessments are done by collecting and analyzing data, while taking into account the variability of hazards and the socio-economic vulnerabilities due to urbanization, rural land use change, environmental degradation and climate change.

Monitoring and Warning Service
This element aims at providing the necessary information. Warning services must have a sound scientific basis for predicting and forecasting and must be reliable enough to operate continuously. This ensures accurate warnings in time to allow action. Warning services for different hazards should be coordinated where possible to gain the benefit of shared institutional, procedural and communication networks.

Dissemination and Communication
This element aims at informing individuals and communities about risks and actions. To be effective, warnings must reach the individuals and communities at risk. This means also that warnings should contain clear, useful information leading to proper responses. Communication channels and tools must be identified beforehand and established at regional, national and community levels. To ensure full dissemination of warnings, the use of multiple and coherent communication channels is necessary.

Response Capability

This element aims at improving the capability to respond to hazards. Communities' education and preparedness programmes should be established to ensure that the proper response and action is undertaken by the individuals and communities at risk at the right time. Disaster management plans should be operational, well practiced and tested.

References

Aureli A, Ganoulis J, Margat J (2008) Groundwater resources in the Mediterranean region: importance, uses and sharing. In: Med 2008 Dossier, IEMED, Barcelona, pp 96–104. http://www.iemed.org/anuari/2008/aarticles/EN96.pdf. Accessed 20 Dec 2017

Davy L (1989) Une catastrophe naturelle: l'averse Nimoise du 3 Oct. 1988 et ses consequences. Hydrol Cont 4(2):75–92

Do ÓA (2012) Drought planning and management in transboundary river basins: the case of the Iberian Guadiana. Water Policy 14:784–799

Ganoulis J (1994) Flood retention basins in the Mediterranean urban areas. In: Rossi G et al (eds) Coping with floods. NATO ASI Series. Kluwer Academic, Dordrecht, pp 759–765

Ganoulis J (1995) Floodplain protection and management in Karst areas. In: Gardiner J et al (eds) Defense from floods and floodplain management. NATO ASI Series, vol 299. Kluwer Academic, Dordrecht, pp 419–428

Ganoulis J (2003) Risk-based floodplain management: a case study from Greece. Intl J River Basin Manag 1(1):41–47

Ganoulis J (2009) Risk analysis of water pollution, 2nd, revised and expanded edition. Wiley VCH, Weinheim, 306 pp

Ganoulis J, Vafiadis M (1995) Urban flood control in Karst areas: the case of Rethymnon (Greece). In: Gardiner J et al (eds) Defence from floods and floodplain management. NATO ASI Series, vol 299. Kluwer Academic, Dordrecht, pp 181–192

Ganoulis J et al (eds) (2011) Transboundary water resources management: a multidisciplinary approach. Wiley-VCH, Weinheim

Gardiner J et al (1995) Defence from floods and floodplain management. NATO ASI Series E, vol 299. Kluwer Academic, Dordrecht

GWSP Digital Water Atlas (2008a) Map 36: mean annual precipitation 1950–2000 (V1.0). http://atlas.gwsp.org

GWSP Digital Water Atlas (2008b) Map 38: mean annual surface runoff 1950–2000 (V1.0). http://atlas.gwsp.org

Hadley Centre (2005) Climate change and the greenhouse effect: a briefing, Technical note 62

IPCC (2001) Climate change 2001. Synthesis report: the scientific basis. International Panel on Climate Change. Cambridge University Press, Cambridge

ISDR (2006) International strategy for disaster reduction. EWC III-third international conference on early warning: developing early warning systems-a checklist. http://www.unisdr.org/2006/ppew/info-resources/ewc3_website/. Accessed 20 Dec 2017

Merrey J (2009) African models for transnational river basin organisations in Africa: an unexplored dimension. Water Altern 2(2):183–204

Molle F et al (2007) River basin development and management. In: Molden D (ed) Water for food, water for life: a comprehensive assessment of water management in agriculture. Earthscan, London, pp 585–625

Rossi G et al (1994) Coping with floods. NATO ASI Series E, vol 257. Kluwer Academic, Dordrecht

Serra P (2000) The defense of Portuguese interests in the agreement on cooperation for the protection and sustainable exploitation of the waters of Luso-Spanish catchment areas. In: Vlachos E, Nunes Correia F (eds) Shared water systems and transboundary issues with special emphasis on the Iberian Peninsula. Luso-American Foundation, Lisbon, pp 229–260

Skoulikaris C (2009) Mathematical modelling applied to the sustainable management of water resources projects at a River Basin scale: the case of the Mesta-Nestos. Joint PhD thesis: Aristotle University of Thessaloniki, Greece and Mines ParisTech, Paris

UNECE (2006) Guidance on water and adaptation to climate change. http://www.unece.org/index.php?id=11658. Accessed 20 Dec 2017

UNESCO (1979) Map of the world distribution of Arid regions. MAB Technical notes 7, Paris, France

Videira N (2007) Integrated evaluation for sustainable river basin governance. In: Kallis G, Videira N, Antunes P, Santos R (eds) Integrated deliberative decision processes for water resources planning and evaluation. IWA Publishing, London, pp 132–148

WRI (1990) World resources 1990–1991. World Resources Institute, Washington, DC

Zeitoun M, Warner J (2006) Hydro-hegemony: a framework for analysis of transboundary water conflicts. Water Policy 8(5):435–460

Chapter 3
Transboundary Water Conflicts and Cooperation

Whiskey is for drinking; water is for fighting over.
Mark Twain, Author

Abstract In this chapter, different types of conflicts and cooperation activities are analysed as related to water quantity, water quality, water ecosystems and socio-economic issues. Conflicts and cooperation can be developed between countries but also between regions, water districts and municipalities located in the same country. International and national water disputes and conflicts are very much related to different political factors and in some extreme cases may lead to civil wars or to armed confrontation between countries. Well known examples of potential transboundary water conflicts and cooperation achievements are, on the one hand the water conflicts between Israel and its neighbours, the water disputes in the Nile basin, the Pakistan-India issues over the Indus River and the Tigris/Euphrates rivers disputes between Turkey, Syria and Irak and, on the other hand, successful examples of cooperation exist, such as the Danube and the Rhine river basins in Europe and the Great Lakes between US and Canada.

The main objective of this chapter is twofold: first, to recognise water districts at risk of conflict, to assess and classify different causes and various types of transboundary conflicts; second, to investigate conflict prevention strategies and the ways to reverse potential conflicts into cooperation, also mentioning the paradoxical existence of cooperative situations without any political agreement. Different tools for conflict management and conflict resolution refer to the so called *water diplomacy* or *hydro-diplomacy*. This includes the international law and regulatory instruments as well as economic and institutional tools, such as the international water authorities and river basin commissions.

Keywords Water conflict · Cooperation · Hydro-diplomacy · Negotiation · Mediation · Resolution

J. Ganoulis, J. Fried, *Transboundary Hydro-Governance*,
https://doi.org/10.1007/978-3-319-78625-4_3

3.1 Assessing Water Conflicts

3.1.1 Shared Waters at Risk

Different causes related to environmental conditions or to water use in various economic sectors, such as domestic water supply and mainly agriculture, can generate water risks at different transboundary scales, such as local and/or regional within the same country, or international between different countries. In this first part of the chapter we concentrate on assessing globally the transboundary waters at risk, either surface or groundwater. If we generally define the risk as a possible negative impact or loss, then we could assess the transboundary waters that are vulnerable to various risks which may range from disagreements to conflicts, disputes and even more to armed confrontations between riparian parties.

The origin of these conflicts may refer to some historical events, fights over territories and national sovereignty issues. A typical example of a classic *upstream-downstream conflict* is the case of the *Indus River Basin*. Originating from the Tibetan Plateau in China and running through almost 3000 km across India and Pakistan down to the Arabian Sea, the Indus River is very important for agriculture and food production in one of the world's largest and fastest growing populations in the world. The water variability in the river basin is very large, the mean annual rainfall ranging from less than 100 mm in the Arabian Sea region to more than 750 mm in the northern Kashmir, below the Himalayas Mountains. Seasonal monsoons exacerbate the vast variability of the river flow, with about ¾ of the flow occurring during 4 months of the year—June to September. Mistrust and hostility between the two countries became international in 1947 with the Partition of British Colonial India dividing the river catchment into an upstream and downstream part and creating two independent dominions, India and Pakistan. Commercial confrontation between both countries around export duties, different religious and socio-economic conditions and also military hostilities over the territorially disputed Kashmir region resulted to India's unilateralism (1948) that discontinued the delivery of water downstream by claiming upstream state sovereign rights. Pakistan protested calling to its human right for water supply and asked for arbitration via the International Court of Justice. India accepted the Pakistan proposal for World Bank mediation services and an agreement was concluded in 1960, known as the "Indus Water Treaty". That allowed to develop an enormous irrigation plan in India that originated a "green revolution" for the next few decades. However, the lack of efficient water storages and distribution techniques that could sustain the expansion of the irrigation acreage, and the extensive use of dams and canals to provide hydropower, dried a large extension of the Indus River, causing displacement of people and deteriorating the socio economic development of the region, as well as the degradation of the Indus plain ecosystem. Nowadays, India has requested to renegotiate the Indus Water Treaty, putting under risk the stability and wellbeing of the region.

Another example of upstream-downstream conflicts is the *Nile River*, which with a total length of 6695 km is the longest river in the world and collects water from 11 African countries before reaching the Mediterranean Sea. Its two main tributaries are the White Nile flowing from the equatorial lake region and the Blue Nile originating in the Ethiopia Highlands. The two tributaries converge in Khartoum, the capital of Sudan where they form the Nile River. Sudan contains the largest part of the Nile watershed. As the Nile riparians progressively gained independence from Colonial powers, riparian disputes became international and consequently more contentious, particularly between Egypt and Ethiopia. The core question of historic versus sovereign water rights is complicated by the technical question of where the river ought to be best controlled—upstream or downstream by existing dams, as the Aswan Dam in Egypt, or new dams, as the Renaissance Dam in Ethiopia.

Apart *sovereignty issues* and the upstream-downstream dissymmetry, *water scarcity* of shared *groundwater resources* is another major cause of conflicts. The *Jordan River Basin* which is internationally shared in the Middle East by five countries: Lebanon, Syria, Jordan, Israel and the Palestinian Authority Territory is a well-known case, where the arid climate is one of the main risk factors for water disputes. Since water is a scarce resource to every country in the region, access to enough water and use of groundwater resources to meet the demand of households and economic sectors, mainly agriculture, is a major concern for all governments involved. In combination with existing regional tensions, which are part of the Israeli-Arab conflict, competition over scarce water resources leads to international political and security consequences as access to water not only touches state sovereignty and integrity, but also issues related to ideology and nationalism.

Wolf et al. (2003) have analysed different biophysical, geopolitical, and socio-economic parameters in order to identify the most important *risk indicators* of water disputes. By developing the Transboundary Freshwater Dispute Database at Oregon State University (www.transboundarywaters.orst.edu) they explored different physical and social variables as causes of possible water disputes. They found that severe conflicts happen historically in places with highly variable hydrologic conditions. Water scarcity, intensive water use in sectors like agriculture and also water quality degradation are all risk indicators of regional political stability. Furthermore, two main results of their work may be summarized as follows:

1. "*internationalization*" of river basins is a significant risk indicator of water disputes. Following the break-up of empires, like the British Empire in the 1940s and the USSR in the late 1980s, the Jordan, Nile, Tigris-Euphrates, Indus, and Aral Sea are typical examples of such basins at risk of conflict.
2. existence/absence of *cooperative transboundary institutions* such as River Commissions or River Basin Organisations is a strong indicator of cooperation capacity or, in the opposite case, of water related disputes. Common riparian institutions are important vectors of political cooperation between riparian countries and the potential for improving common technical issues at the transboundary watershed scale.

By using these two risk indicators i.e. basins with/without institutional capacity and internationalized basins, Wolf et al. (2003) identified basins at risk of potential conflict around the world. In terms of potential conflicts and/or institutional capacity, these basins include: the Ganges-Brahmaputra, Han, Incomati, Kunene, Kura-Araks, Lake Chad, La Plata, Lempa, Limpopo, Mekong, Ob (Ertis), Okavango, Orange, Salween, Senegal, Tumen, and Zambezi. These basins at risk are shown in Fig. 3.1, where also those basins in conflict at that time or with active negotiations have been added, i.e. the Aral, Jordan, Nile, and Tigris-Euphrates basins.

We should notice that the map shown in Fig. 3.1, based on particular indicators, represents a picture taken during a certain time period. The situation may change dynamically depending on other water related socio-economic and political factors that may lead to different risks. Basins at risk may change on time depending on the severity degree of the socio-economic and political situations. In many circumstances, water related causes of conflict are hidden behind more visible socio-economic parameters and there is need to carefully analyse every particular case in order to recognise the water related roots of conflict.

Two recent examples of hidden water-related causes of conflict that need to be decrypted are the so-called "Arab Spring" and the present "Syrian conflict and civil war" in the Middle East. In 2008 and also during 2010–2011 severe droughts in Canada, USSR and Argentina combined with extreme wild-fires in the former Soviet Union resulted in dramatic loss of cereal production followed by considerable increase of cereal prices in the international stock market. Arab countries import 50% of food (World Bank 2008) and as a consequence the price of bread, normally subsidised by Arab authoritarian regimes in countries like Tunisia and Egypt, has been increased substantially. In 2010, in the rural town of Sidi Bouzid, Tunisia, a young man who was breadwinner for his widowed mother and six brothers, set himself on fire in front of a government building as an act of desperation not having

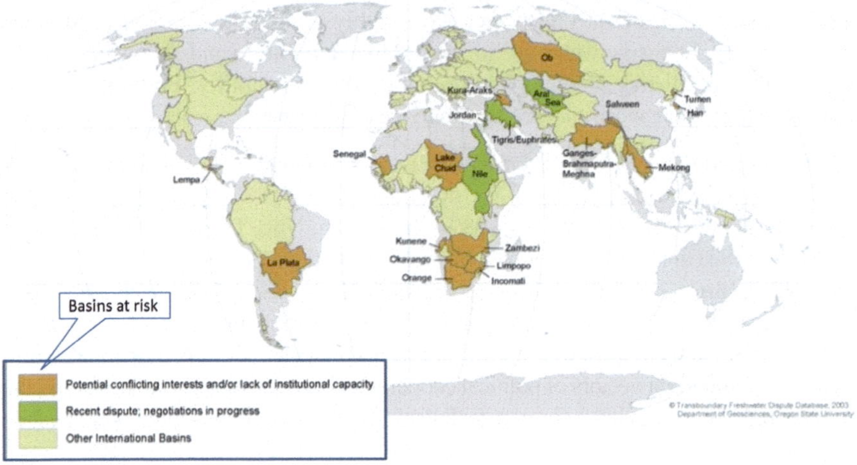

Fig. 3.1 Basins at risk (Wolf et al. 2003)

the possibility to sell goods for feeding his family. His act captured by cell phone cameras and shared on the Internet produced within days, public protests across the country, calling upon President Ben Ali and his regime to step down. About a month later, the President fled.

Since 2003, when the Iraq war begun, about 4.2 million refugees have fled Syria in all directions and 7.6 million people moved in Syria as Internally Displaced Persons (IDPs). War and complex socio-economic and political factors were some of the drivers for these refugees fluxes. Furthermore, some new findings published in the Journal of Geophysical Research-Atmospheres (Cook et al. 2015) indicate that a series of severe droughts starting in the late 1990s in Syria were at the origin of the mass migration of rural farmers into Syrian cities (Fig. 3.2). By analysing tree ring data that covered 900 years of drought history the study suggests that apart previous repeated droughts, the prolonged and devastating 2005–2010 drought is likely the worst to affect the region in 900 years and contributed to the 2011 Syrian uprising. Although natural variability and droughts are common in Eastern Mediterranean and the Middle East at decadal and multi-decadal scales, these exceptional repeated droughts indicate that a probable climate change signal is modifying the water cycle in the region (Kelley et al. 2015).

The Syrian example illustrated in Fig. 3.2, strongly demonstrates that *water governance* is, at top of other natural and anthropogenic factors, the most crucial indicator for water related socio-economic risks. In fact, when in 1971 Hafez al-Assad became President, he initiated, without any concern to sustainable use of land and water resources, an ambitious programme for making Syria self-sufficient

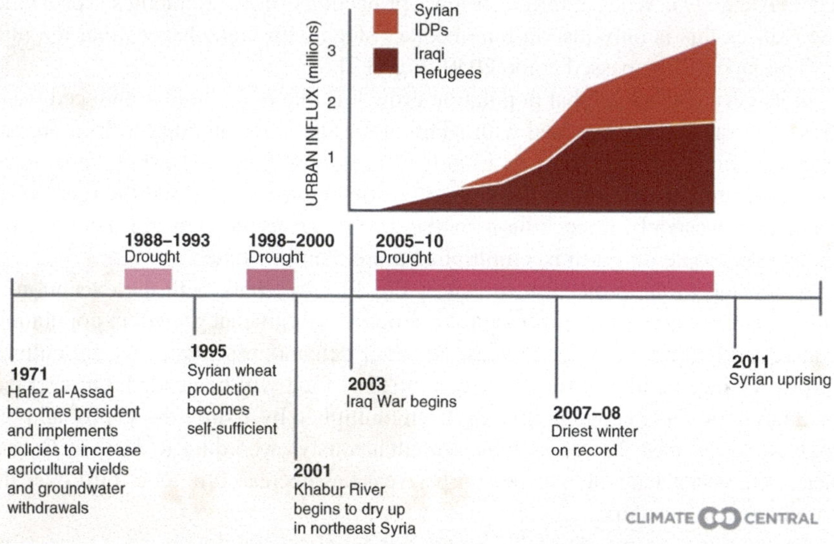

Fig. 3.2 Agricultural policy and droughts as risk factors in the Syrian conflict and civil war (from Climate Central 2015 and Kelley et al. 2015)

in cereal based food. As a consequence, due to rainfall decrease in the 6-months winter periods during the repeated droughts, the surface water resources from irrigation canals fed by river tributaries have been depleted. Farmers not having access to surface water have pumped groundwater in quantities largely exceeding the renewable groundwater resources.

Although, as shown in Fig. 3.2, the country became in 1995 self-sufficient in wheat production, 6 years later, in 2001, the Khabur River started to dry. The Khabur River originating from Turkey is a tributary of the Euphrates transboundary river and although usually an ephemeral river, it is an important source for irrigation water in the North-Eastern Syria. Its recent disappearance is strongly dependent on the depletion of groundwater resources in the region due to agricultural over-use. The government attempted to control the groundwater depletion by publishing a law in 2005 requiring a license to dig wells, but the legislation was not implemented. The depletion of groundwater during the recent drought is clearly shown from remotely sensed data by the NASA Gravity Recovery and Climate Experiment (GRACE) Tellus project (Kelley et al. 2015).

The global increase of population on Earth is an important factor for putting water availability and water related conflicts at risk. The exact number of people living on Earth in the coming decades is uncertain, but we do know that the population will continue to grow, and this will impact water availability. Investing in efforts that slow the rates of population growth through increasing access to voluntary family planning services can help ensure that sufficient water is available for global food production, ecosystem health, and political and social stability. The year 2011 will remain as a milestone in the history of human evolution. On 31st October 2011, the total global population reached the symbolic and record figure of seven billion (UNITAR 2017). When referred to most projections of demographics worldwide, one realises this is only just an intermediary stage with high chances that the nine billion mark will be passed come 2050 (Fig. 3.3).

In its very essence, global population growth is one major human-induced water stressor because it is correlated with additional water consumption for drinking and other economic activities. For thousands of years, until the mid-twentieth century, the planet only counted up to a maximum of two billion people and the total world population was only three billion in the 1960s. Over the past half-century, the number of people on Earth has multiplied more than two times.

According to the projection shown in Fig. 3.3, by 2050, with the exception of Europe, every continent is expected to experience a significant growth in population. Consequently, this means an increase in water demand, essentially for agricultural purposes, since continuously larger quantities of food will be needed. On average, the amount of water consumption has been multiplied by 7 since the 1900s, and has doubled in the past 20 years alone. Simultaneously, according to UNESCO estimates, the water available per person per year has decreased by one-third over the last quarter of a century.

The massive increase in global population numbers and the growth of emerging countries which more than ever require large quantities of freshwater to develop their economies and enable entire communities to increase their level of development

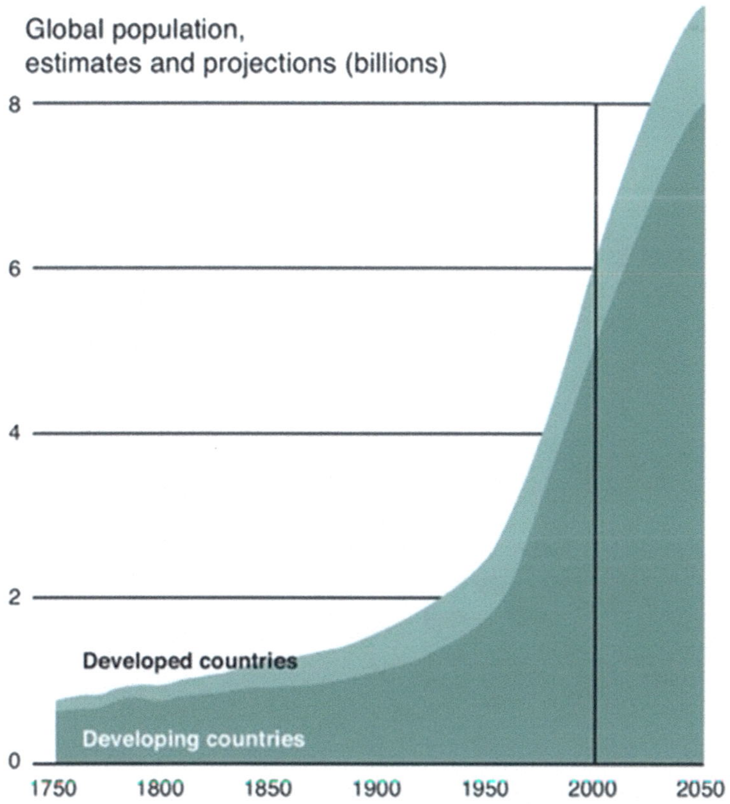

Fig. 3.3 Global population trends (UNEP/GRID-Arendal Maps and Graphics Library 2017)

resulted in an exponential increase of water demand (UNITAR 2017). Per continent and per year during the time period 1900–2025, this increase of water withdrawals is shown in Fig. 3.4, where the water demand is expressed in km^3/year or 10^9 m^3/year or billions of m^3/year.

Population density is the determining factor in setting the amount of water theoretically available for each person. For example, while Asia's rivers, which are the primary source of freshwater for a majority of people, have the highest water flow rate in the world, each inhabitant in Asia has the least amount of water to drink or use because of the high levels of population density. Limited amounts of water for a high number of individuals is a phenomenon often referred to as "water crowding": an increasing pressure is being placed on finite, erratically available and vulnerable water resources as population rises. Concentration of population is mainly observed in cities, coastal areas and river's deltas.

We have previously enumerated different qualitative indicators of risk for future water conflicts. In quantitative terms, the most widely cited quantitative indicator for water resources management is the Malin Falkenmark's Water Stress Indicator

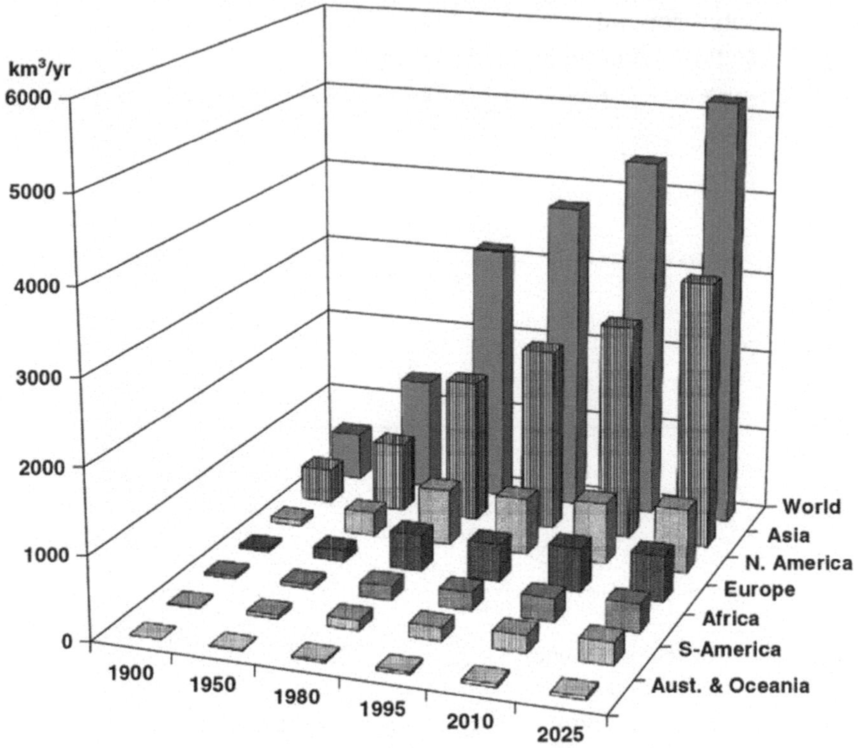

Fig. 3.4 Water withdrawals per continent from 1990 to 2025 (Shiklomanov and Rodda 2003)

(1989). This is obtained by dividing the annual volume of available renewable water resources of each country by its population. Various levels of "water stress" may be defined as follows:

Annual renewable water per capita (m^3/person)	Water stress indicator = Country's annual renewable water/population
> 1700	Water security
1000–1700	Periodic water stress/vulnerability
500–1000	Chronic water stress/stress
< 500	Absolute water stress/scarcity

As shown in Fig. 3.5 many Mediterranean countries experience already chronic and absolute water stresses and the situation will become more critical in the future. In fact, without taking into account possible reduction of available renewable water resources due to climate change, the decrease of the indicator values in 2025 and 2050 shown in Fig. 3.5 is due to the population increase.

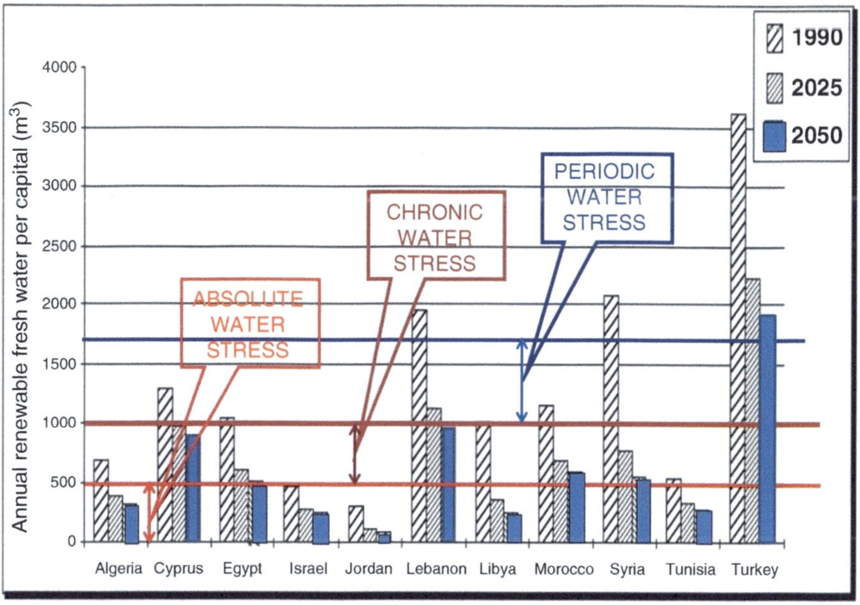

Fig. 3.5 Actual and projected Water Stress Indicator in selected Mediterranean countries

Though commonly used, Falkenmark's indicator has been criticized on a number of grounds, mostly that it accounts neither for spatial variability in water resources within countries, nor for the technological or economic adaptability of nations at different levels of development.

However, as shown in Fig. 3.6 this indicator can be applied regionally and/or at the river's watershed scale, when data on renewable water and population are available. Values of the ratio expressing the Water Stress Indicator are low in the South and Eastern Mediterranean coastal areas (MENA region), because of low water availability (value in the nominator) and high amount of population (value in the denominator).

To improve Water Stress Index, Ohlsson (1999) developed a "Social Water Stress Index," which incorporates "adaptive capacity" into Falkenmark's measure, essentially weighting the index by a factor based on UNDP's Human Development Index. While Ohlsson's index is a useful contribution, it also misses the spatial component. Similarly, neither Falkenmark nor Ohlsson suggest much about the geopolitical results of scarcity, focusing instead on implications for water management of water availability per person.

Peter Gleick (1993), suggested four vulnerability indices to determine "regions at risk" for international water conflicts. These are:

1. ratio of water demand to supply,
2. water availability per person (Falkenmark's water stress indicator),
3. fraction of water supply originating outside a nation's borders, and
4. dependence on hydroelectricity as a fraction of total electrical supply.

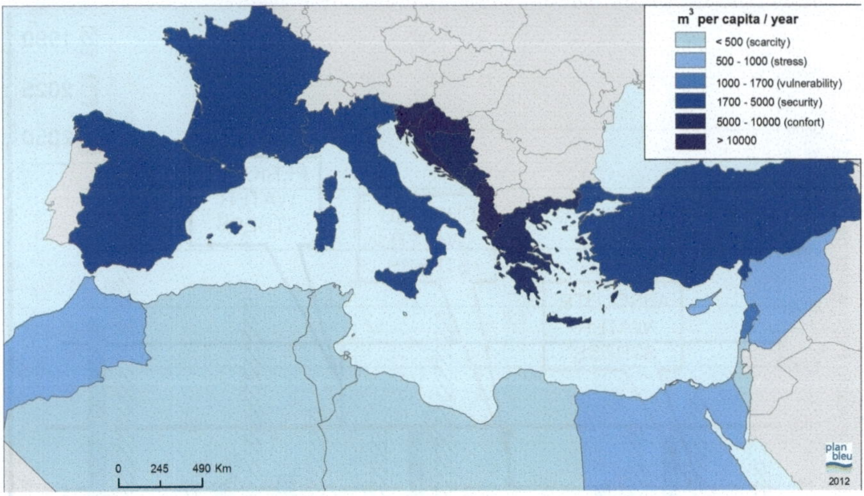

Fig. 3.6 Distribution of the Water Stress Indicator along the Mediterranean watersheds (Plan Bleu 2012)

Gleick's indices, like Falkenmark's and Ohlsson's, focus on the nation as the unit of analysis and on physical components of water and energy. Moreover, as countries develop, rising standards of living also boost the demand for modern automated appliances (dishwashers, washing machines etc.), non-food items and services, many of which have a large ecological water footprint, which is the amount of water used to produce the goods and services we use.

Whereas in developed countries, consumers and corporations are increasingly being encouraged to help reduce this footprint, large amounts of water are used to produce and process non-food goods and services, adding more pressure on quantitative and qualitative standards of water resources.

The concept of virtual water has precisely been introduced to reflect the indirect use of water that leaves a country in exports and enters another through imports as integral part of goods that use water for production: introduced by Prof. Tony Allan, the virtual water content of a product (a commodity, good or service) has been defined by Hoekstra and Chapagain as "the volume of freshwater used to produce the product, measured at the place where the product was actually produced". It refers to the sum of the water use in the various steps of the production chain (https://en.wikipedia.org/wiki/Virtual_water). Another way to think of possible transboundary competition for water is to consider the extent to which some countries are dependent on neighbouring countries as their source of water. Figure 2.7 "Distribution of the *dependency ratio* between shared and national water resources" in Chap. 2 illustrates this current state of affairs.

Within national borders, the limited water supply, notably in arid and semi-arid regions, poses an increasing risk to societies' stability as they are faced with either one or several of the following stressors: strong population growth (true in most

African countries and India), developing economies on the rise (e.g. BRICS, acronym for Brazil, Russia, India, China and South Africa), and climate change, probably the most evenly distributed threat among nations most prone to water scarcity.

It is a worth repeating statistic that "one third of the world's population today lives in water-stressed countries". By 2025, projections show that two thirds of the total population will be at risk of finding itself confronted with severe cases of water scarcity. More than ever before, water is a resource that requires to be shared evenly, managed reasonably, consumed parsimoniously, and preserved carefully. Any excess, either due to man's influence or simply a cause of environmental influxes, brings us closer to the possibility of a dispute breaking out.

There have of course always been shortages of water in some places on Earth at some times. Whenever this happened where there were people, there was competition for water, and sometimes disputes. But humans always learned to adapt or cope, sometimes by relocating in less hostile regions. At first people followed the water, settling near rivers, lakes, and springs, and moved to others if these dried up because of climate variability. As technology evolved, people became able to move the water to them by building reservoirs, aqueducts, and pumping stations.

However, in the past hundred years, our population has mushroomed and large cities and megacities have developed. Water consumption has risen to feed us, quench our thirst, and supply the industries that feed our economic growth. Pollution loads have outstripped the capacity of our ecosystems to respond. Locally and regionally, competition for water is increasing. To this must be added the threats to regional and global ecosystems caused by anthropogenic and natural climate change.

3.1.2 Conflict Identification

Conflicts may have different intonations and various degrees of severity. We may agree that the terms (1) conflict, (2) dispute, (3) armed conflict and (4) war may be used in ascending order of severity degree.

Identifying a water conflict is the first step for choosing the most appropriate mechanism for conflict resolution. In a water related conflict two main components are distinguished: (1) diagnostic analysis of the particular characteristics of the conflict and (2) recognition and study of the potential stressors that may cause the conflict. The aim is to understand better the conflict and other related issues by analysing them scientifically (UNITAR 2017).

Conflicts over water could be looked upon as consisting of three key spheres: the hydrosphere (natural), the socioeconomic, and the political (both anthropogenic) (Fig. 3.7).

In the hydrosphere, strong potential conflict may exist on water quantity and quality between human and ecosystem's needs for water and needs related to economic development. Within the context of human needs, conflicts over water generated within the water sphere itself are often affected by problems and disputes

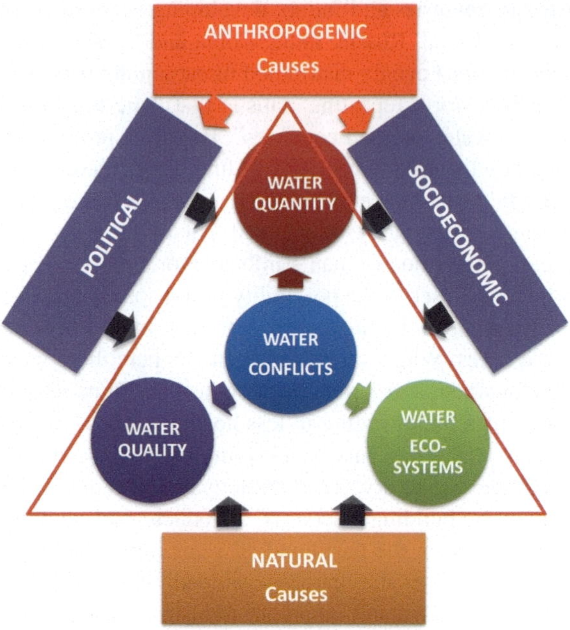

Fig. 3.7 Classification of water conflicts in three spheres: Natural, Socio-economic and Political

in the other two spheres. This has led some to claim that water wars are inevitable. However, a counter-movement claims that there is room for peace to be built around water, which can be achieved by fostering cooperation between nations so as to share the benefits stemming from water. Supporting this idea, the World Water Vision report published early in 2000 concluded that the planet does not suffer from a shortage of water, but instead that the precious resource is more a victim of "*mismanagement or inefficient water governance*" (UNITAR 2017). Water governance or hydro-governance is the main topic of this book to be addressed in the following two parts. Beforehand, definitions of water disputes and water conflicts, their causes and their classification are given below (UNITAR 2017).

Defining a Water Dispute (from UNITAR 2017)

Disputes, it should be noted, always involve at least two parties, each of which is trying to do a good job of meeting its own objectives. By the nature of the situation or circumstances, the two or more parties come to see each other as obstacles to reaching their respective goals.

Unless there is some sort of intervention, a dispute may grow to the point that the parties come to see each other as adversaries, even as enemies. Communication becomes distorted, people view each other as stereotypes, no longer as human beings. Each new escalation in aggressive behaviour is justified as a counter-response to the other person's perceived aggression. When this kind of polarisation occurs, most assume that they are now in a struggle to "win", even if it means that the other person should "lose". This results in a "win-lose" battle or, as it is sometimes called, a "zero-sum game", meaning that everything you gain, whether it is money, status, power, authority, must necessarily be at someone else's expense, or vice versa.

When this kind of dispute occurs, it is usually dysfunctional, whether within or between organisations or among nations and across water sectors. It can prevent people from working together even when they share common goals and can cause such anger and stress that the relationship is destroyed, even though it has been and could continue to be of benefit to the parties. In effect, disputes consume invaluable amounts of time and resources needed for more productive projects.

Defining a Water Conflict (from UNITAR 2017)
First of all, taking apart the term "conflict" should enable us to analyse what are the terms and what makes up a "water conflict". For the sake of a broader understanding of the sole term "conflict" let us look into some of the definitions found in the dictionary and specialised literature. The Oxford Dictionary offers the following, identifying conflict as:

(1) An encounter with arms; (2) A fight; (3) A conflict of interests; (4) An incompatibility between parties.

A conflict may be seen solely as a state of incompatibility between parties and their claims, with the only risk of this evolving into an actual dispute depending on the psychological perception of this proven incompatibility. In other words, conflicts may remain hidden if the other parties do not realise that they are negatively affected or can be open once they have realised it and do not accept the situation.

One important thing is to understand that the word conflict has many connotations, ranging from a state of open prolonged fighting to a situation where two or more entities (groups of people, nations or states) wish to carry out acts which are mutually inconsistent. We shall therefore agree here that conflict is not necessarily equal to war or a state of effective armed fighting between parties, but rather a state of incompatible aspirations between parties around the use of water.

Causes of Water Conflicts (from UNITAR 2017)
Water in parts of the world, as a fundamentally scarce resource, has long been a conflict-causer between people, communities, and nations. These conflicts can arise nationally and internationally and are due to the competing needs to utilise the water for a variety of purposes: household requirements, irrigation, fishing, industry, hydroelectricity, etc. Arising from a range of issues, conflicts can occur between riparians situated upstream—who control the sources—and those downstream, who are at a disadvantage (i.e. the Nile River, between Sudan and Egypt), between those on the two banks of a river (i.e. the Parana River, between Paraguay and Brazil), or around a lake (i.e. Lake Chad, See of Galilee).

The competing needs cited above can be grouped under broad categories which are usually found to be those which trigger conflicts. Hereunder, different domains or water conflict triggers are described and illustrated.

Water Quantity
Concerns over quantity arise when the resource is not sufficient to meet all the needs of the riparian countries. A classic scheme is when a country upstream, which controls the source, claims sovereignty over the water, resulting in a conflict with the downstream countries. Turkey, for instance, claims sovereignty over the

Euphrates and Tigris rivers, and started the GAP project, which will be beneficial to Turkey and a potential threat to downstream countries. This is notably a source of concern to Syria and Iraq, and has resulted in alleged tension with Turkey.

Water Quality
When upstream countries carry out activities that have a degrading impact on the overall quality of water (i.e. pollution as generated by chemicals disposed by factories or ships), this can end in a situation of conflict with downstream water users. Many countries along its route use the river Rhine. Pollution of the river has been caused by the chemical industries of Germany, Switzerland, and France, as well as the shipping industry along the river. The shipping issue was resolved in 1950 by creating an International Commission. The chemical pollution imposed a burden on the Netherlands, who uses the water mainly for drinking and agricultural purposes. The case was brought to the European Court of Justice, while the conflict was aggravated by serious chemical accidents, which resulted in the 1987 and 1991 Chemical Agreement.

Flooding
Upstream countries or landowners influence downstream flooding by clearing their land of trees and natural vegetation, urban development, building drainage works, training the river, and building dikes. This augments and accelerates the flow in the river, which increases the probability and severity of downstream flooding.

Hydroelectricity Production
Hydroelectric production requires for dams to be constructed, and building such infrastructures usually clashes with other uses of water. Indeed, dams need to be as full as possible and only release the amount of water necessary to produce the needed amount of electricity. Therefore, the operating strategy of a hydroelectric dam is often incompatible with water supply, crops irrigation, fish requirements, ecological considerations, and hydroelectric projects in a given country can potentially lead to conflict with downstream riparians. Actually, India, Nepal, and Bangladesh are still at conflict over how to use the water of the Ganges-Brahmaputra. India and Nepal upstream have an interest in the hydroelectric power potential, while Bangladesh has an interest in flood control, and concerns over water shortage.

Environment/Ecology
Conflicts between states can also arise from environmental and ecological issues related to water. Two interesting examples corroborate this point.

First, the conflict that erupted between the United States and Mexico in the 1970s was over the increase of the nitrate content and the salinity of the Colorado caused by the United States. It was resolved in 1973 with a joint protocol, the United States agreeing to build the Yuma plant in order to reduce the salinity of the water and help it reach an acceptable quality.

Secondly, a conflict still underway between China and Russia relates to ecological deterioration and environmental problems in Russia. China, which is the upstream country on the Tumen River, does not have wastewater treatment plants and is polluting the river to such an extent that the fish in the river contain a high level of chemicals that cause cancer. The water in the lower Tumen River is so contaminated that it is not suitable for municipal or industrial use. This situation has increased the tension between the two countries, and seems to have gone beyond the environmental and ecological issues, causing some security issues to be raised between them.

Economic Issues

Excessive use of water on the basis of economic development can create dire situations for all users, with a non-negligible risk for conflict and extensive and irreversible harm made to the environment. The Gabcikovo-Nagymaros Dam project for example was proposed in order to provide for electricity, flood control, and improved navigation on the Danube for Hungary and Czechoslovakia. The initiative resulted in the 1977 bilateral treaty. Things remained stable until 1988 when Hungary withdrew from the project because of internal political pressures and its environmental concerns. Slovakia, which was the natural successor to the rights and obligations of Czechoslovakia, continued the project and constructed a channel to divert water of the Danube from Hungary. This project was economically important for Slovakia to produce much needed electricity. In 1992, Hungary took Slovakia to the International Court in The Hague, claiming environmental damages for its unilateral action in diverting the Danube. The court did not make a definitive ruling but rather urged the parties to reach an agreed solution.

Another example concerns the case of the Aral Sea. Until 1960, the Aral Sea was the fourth largest lake in the world. Excessive use of the water of the two rivers that feed this inland lake—the Amu Darya and Syr Darya—and of the lake itself for irrigation in the time of the former Soviet Union created an ecological and economic disaster. The fishing industry disappeared, salinity harmed the crops, water was polluted with pesticides, and the Aral Sea was reduced to about a quarter of its natural size. This has also had an adverse impact on human health and mortality, and caused desertification of the region. The five Central Asian Republics that share the water of the rivers and the lake are trying to deal with these issues and cooperate in jointly managing the transboundary water resources.

According to the previous experience developed by the UNESCO Chair/INWEB (International Network of Water/Environment Centres for the Balkans: www.inweb.gr), there are many types of water conflicts which are the origin of obstacles for implementing effectively the Integrated Water Resources Management (IWRM) framework on internationally shared waters.

Conflicts may arise from the following causes:

- lack of political willingness for cooperation between countries
- differences in socio-economic and cultural level
- lack of trust between scientists and policy makers
- different countries interests
- conflicting objectives
- different priorities between countries
- sovereignty and possible territorial claims

- differences in the use of technical standards and specifications for data collection and information sharing
- lack of harmonisation in methodological approaches involving conceptual and analytical modelling of hydrological, environmental and socio-economic processes

Hydro-governance through good cooperation between riparian countries is the best way for addressing such water conflicts.

3.1.3 *Typology of Conflicts*

Water conflicts can be classified into different types according to the:

1. Degree of intensity;
2. Spatial extent; and
3. Conflict's Risk Origin.

(1) Degree of Conflict Intensity

According to Yoffe et al. (2003), experts in water conflict and cooperation, the intensity of water conflicts may vary *from cooperation to armed conflict*. In the scale presented in Table 3.1, 0 means neutral situation and the scale extends from −7 (war declaration) to +7 (peaceful voluntary unification into one nation). The scale also demonstrates that even the less intense conflicts are still forms of conflict. In other words, the absence of war does not mean the absence of conflict. The scale not only identifies that conflicts vary in intensity towards disputes and wars, but that the same conflict can undergo various degrees of intensity through time.

Table 3.1 The Basins at Risk (BAR) water event intensity scale (Yoffe et al. 2003)

Scale	Event description
7	Voluntary unification into one nation
6	Major strategic alliance (international freshwater treaty)
5	Military, economic or strategic support
4	Non-military economic, technological industrial agreement
3	Cultural or scientific support (non-strategic)
2	Official verbal support of goals, values, or regime
1	Minor official exchanges, talks of policy expressions—mild verbal support
0	Neutral or non-significant acts for the inter-nation situation
−1	Mild verbal expressions displaying discord in interaction
−2	Strong verbal expressions displaying hostility in interaction
−3	Diplomatic—economic hostile actions
−4	Political—military hostile actions
−5	Small scale military acts
−6	Extensive war acts causing death, dislocation or high strategic costs
−7	Formal declaration of war

Table 3.2 Conflict dynamics at different spatial levels (from UNITAR 2017)

Geographic scale	Characteristics
International	• Disputes on transboundary waters can arise between riparians or states themselves. • Very little violence, but existing tensions between parties are all the more pervasive and difficult to overcome, resulting in degraded political relations, inefficient water management, and ecosystem neglect. • There is a long, rich record of conflict resolution and development of resilient joint institutions.
National	• Disputes can arise between subnational political units, including provinces, ethnic or religious groups, or economic sectors. • Higher potential for violence than at international level. • Rationale for international involvement more difficult, given national sovereignty concerns.
Local	• Loss of water-based livelihoods (due to loss of irrigation water of freshwater ecosystems) can lead to politically destabilising migrations to cities or neighbouring countries. • Conducive of local instability which can destabilise regions. • Poverty alleviation is implicitly tied to ameliorating such security concerns.

(2) Spatial Scale of Conflicts

Although challenges on water sharing and managing are essentially reduced to the transboundary scale, it is interesting for comparative purposes to further the analysis of potential water conflicts under their more unusual yet alternative formats which are the International, *National* and *local* levels.

To better help in our contextualisation of the transboundary versus the national and local levels, Table 3.2 below offers a succinct description of conflict dynamics as observed at three different spatial levels at which water conflicts may occur.

(3) Risk Origin and Descriptors of Water Conflicts

The World Resources Institute has developed on the Internet the "Aqueduct Water Risk Atlas", an online mapping tool showing patterns of water stress, flooding, drought, and areas of water access, at the following website: (http://www.wri.org/resources/maps/aqueduct-water-risk-atlas). It combines 12 water risk indicators that one can customise to create maps showing where water-related conflicts might arise.

However, the lack of reliable risk-event data in many parts of the world and the combination methodology of various descriptors into one composite indicator are still in doubt. In fact, such aggregated indicators are neither always objective nor match the reality, as they rather reflect the particular expertise and judgement of the author.

Transboundary water conflicts and potential disputes deserve a dedicated scientific analysis within each international basin, by taking into account reliable data and robust descriptors of natural, ecosystems, human, socio-economic and water governance characteristics. Tensions and conflicts inherent to the international setting further complicate the joint management of watercourses and aquifers as states strive to secure their own freshwater supplies. Putting transboundary water conflicts into context will help gain a clearer view of what is geopolitically conducive to a situation of potentially heightened tensions between two or more states over the claim of freshwater resources.

In this direction it is very useful to learn about a recent project called TWAP (Transboundary Water Assessment Program, website: http://www.geftwap.org/). TWAP is a program of scientific character involving a large number of international partners, Universities and Research Centres, coordinated and sponsored by the Global Environmental Facility (GEF). The main aim of TWAP is to assess the global issues threatening ecosystems and human populations in transboundary water systems and to promote sustainable governance in transboundary water regions. The TWAP contains five transboundary water systems:

1. Groundwater,
2. Lake Basins,
3. River Basins,
4. Large Marine Ecosystems (LMEs),
5. Open Ocean.

TWAP has developed a data portal available at the following address: http://www.geftwap.org/data-portal

The websites below host the results of these five particular TWAP components. Although marine ecosystems and open ocean may be considered as transboundary, the sea being the common boundary for all coastal states, when transboundary freshwater resources are concerned, the most important components are the three first, i.e. the transboundary rivers, lakes and groundwater aquifers.

Water system	Web site	Indicator results
Groundwater	http://www.isarm.org/publications/467	http://twapviewer.un-igrac.org/
Lakes	http://www.ilec.or.jp/en/twap	http://www.geftwap.org/data-portal
River basins	http://twap-rivers.org	http://twap-rivers.org/indicators
Large marine ecosystems and Open ocean	http://onesharedocean.org/	

One of the key outputs of the TWAP River Basin (TWAP RB) component is the delineation of 286 transboundary river basins. From earlier studies, 214 transboundary basins have been identified by UNDESA (1978), 261 by Oregon State University, 263 by OSU (2002), 276 by OSU (2012). The increase to 286 of the number of transboundary river basins is due, on the one hand, to the improvement of the global Digital Elevation Models (DEMs) and, on the other hand, to the breaking up of countries.

The TWAP RB assessment is based on a number of indicators (Table 3.3) and the methodology used is not intended to be a global 'state-of-the-environment' assessment, but rather a relative analysis of basins based on risks to societies and ecosystems. During this design phase, five thematic groups of issues were identified as being of relevance to both populations and ecosystems: (1) water quantity,

Table 3.3 TWAP RB risk indicators of the baseline and the projected risks

Thematic group	Indicator	
Baseline transboundary status (2010)		Projected transboundary stress (2030/2050)
Water quantity	1. Environmental water stress 2. Human water stress 3. Agricultural water stress	1. Environmental water stress 2. Human water stress
Water quality	4. Nutrient pollution 5. Wastewater pollution	3. Nutrient pollution
Ecosystems	6. Wetland disconnectivity[7]Ecosystems 7. Ecosystem impacts from dams 8. Threat to fish 9. Extinction risk	[Environmental water stress]
Governance	10. Legal framework 11. Hydropolitical tension 12. Enabling environment	4. Exacerbating factors to hydropolitical tension
Socioeconomics	13. Economic dependence on water resources 14. Societal wellbeing 15. Exposure to floods and droughts	5. Change in population density
Water systems links		
Lakes	1. Lake influence	
Coastal areas	2. Delta vulnerability (sea level rise, wetlands, population and governance)	

Let us consider the following two socio-economic indicators

(2) water quality, (3) ecosystems, (4) governance, and (5) socio-economics. The indicators assessing the baseline status (2010) and also the "projected risks" for 2030 and 2050 are shown in Table 3.3.

(1) Economic Dependence on Water Resources (No. 13 in Table 3.3)
This indicator measures (1) a fraction of economic activities, (2) the agricultural, (3) the energy-related, and (4) the industrial water withdrawals of all riparian countries that fall within the river basin. It is a measure of the average economic dependence on waters of the countries that share a common river basin. Understanding the degree to which economic activity is concentrated in given basins, and therefore the level of dependence on freshwater resources within basins, will help to illuminate the risk to economies within a basin, should water supplies be substantially altered. This same metric can also help to assess the level of human pressure on water resources.

(2) Societal Wellbeing (No. 14 in Table 3.3)
Low levels of socioeconomic development and human wellbeing put populations at higher risk to low and high flow extremes, and to water pollution. This indicator is composed of five sub-indicators, so the rationale for each is described in turn: (1) Access to improved drinking-water supply, (2) Access to improved sanitation, (3) Adult literacy that indicates the level of education, (4) Infant Mortality Rates (IMRs) that is a useful proxy for overall poverty levels, and (5) Gini coefficients that represent the level of inequality in a basin. Poverty alleviation is implicitly tied to ameliorating such security concerns.

The Figs. 3.8, 3.9 and 3.10 have been drawn using the TWP interactive map, available on the Internet. Concerning the first indicator of economic dependence on water, mainly on agricultural water, we can see that basins like the La Plata in S. America, the Nile in Africa, the Tigris-Euphrates are at economic risk. If we

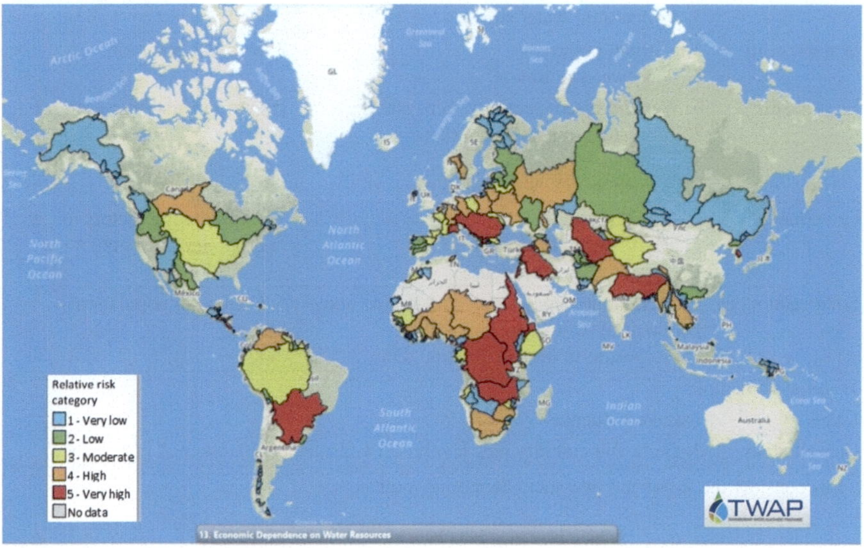

Fig. 3.8 *Economic dependence on water resources:* risk distribution among shared basins

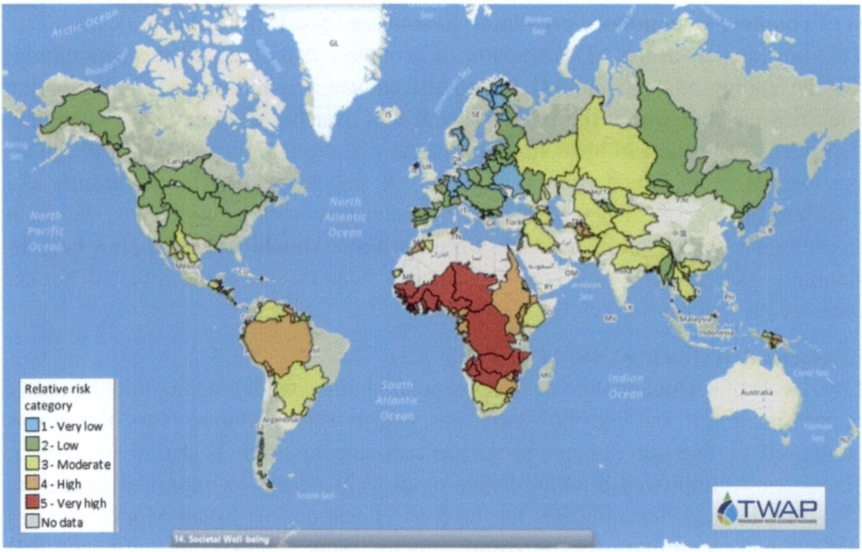

Fig. 3.9 *Societal well being:* risk distribution among shared basins

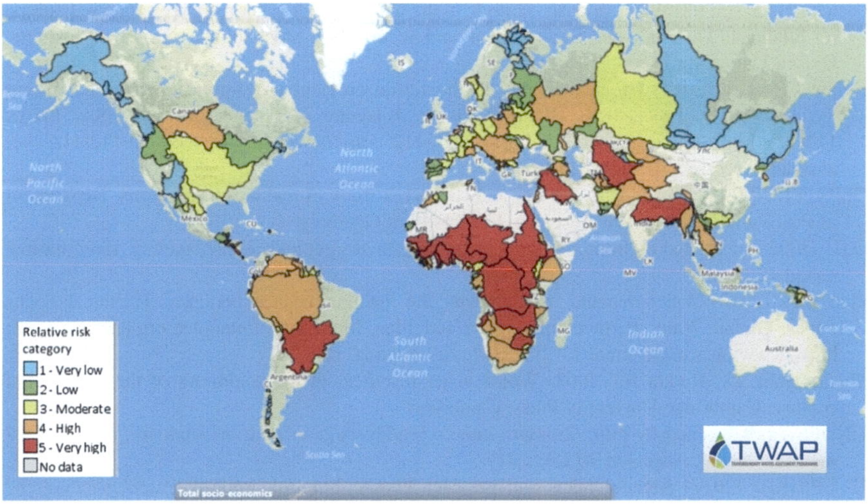

Fig. 3.10 *Composite socio-economic:* risk distribution among shared basins

superimpose the social risk (Fig. 3.9) then we can assess the socio-economic vulnerability of the same river basins (Fig. 3.10).

As indicated in Table 3.3, risks on transboundary river basins are linked to lakes and coastal waters. Lakes are important in providing buffering and storage capacity within transboundary river basins, thus directly influencing water quantity and quality within a given basin. The Lake Influence Indicator aims to highlight these important interactions and interdependencies, focusing on lake buffering and storage capacity within TWAP river basins. Also, many river basins assessed under TWAP RB include deltas, occurring where a river flows into a lake or the sea. The physical geography of deltas often differs markedly from that in the neighbouring parts of the basin, in terms of relief, subsurface characteristics and hydrology. At the same time, many deltas are centres of large populations, agricultural production and economic activity, while maintaining direct connections to the health of the respective river basins. Deltas are therefore given special attention in the TWAP RB, and four Delta Vulnerability Indicators were included for a selected number of deltas (Table 3.3: sea level rise, wetlands, population and governance).

References

Climate Central (2015) http://www.climatecentral.org/search?q=Syria. Accessed 20 Dec 2017

Cook BI et al (2015) Spatiotemporal drought variability in the Mediterranean over the last 900 years. J Geophys Res Atmos 121:2060–2074. https://doi.org/10.1002/2015JD023929.

Falkenmark M (1989) The massive water scarcity now threatening Africa: why isn't it being addressed? Ambio 18:112–118

Gleick PH (ed) (1993) Water in crisis: α guide to the world's fresh water resources. Pacific Institute for Studies in Development, Environment, and Security and Stockholm Environment Institute. Oxford University Press

Kelley CP et al (2015) Climate change in the fertile crescent and implications of the recent Syrian drought. Proceedings of the National Academy of Sciences of the USA, PNAS 112(11). https://doi.org/10.1073/pnas.1421533112. http://www.pnas.org/content/112/11/3241.abstract. Accessed 20 Dec 2017

Ohlsson L (1999) Environment, scarcity and conflict: a study of malthusian concerns. Department of Peace and Development Research. University of Göteborg

OSU (2002/2012) cited by http://ihp-wins.unesco.org/layers/geonode:transboundary_river_basins/metadata_detail

Plan Bleu (2012) Water demand management: the mediterranean experience. Technical Focus Paper. Global Water Partnership (GWP) and Plan Bleu. https://planbleu.org/en/node/1145; http://www.pnas.org/content/112/11/3241. Accessed 20 Dec 2017

Shiklomanov IA, Rodda JC (2003) World water resources at the beginning of the twenty-first century. Cambridge University Press, Cambridge

UNDESA (1978) cited by http://ihp-wins.unesco.org/layers/geonode:transboundary_river_basins/metadata_detail. Accessed 20 Dec 2017

UNEP/GRID (2017) Arendal maps and graphics library. http://old.grida.no/graphicslib/Search?q=Global+population. Accessed 20 Dec 2017

UNITAR (2017) UN Institute for Training and Research, Introduction to Water Diplomacy, Lecture Notes, Module I. https://www.learnatunitar.org/. Assessed 26 Dec 2017

Wolf A et al (2003) Conflict and cooperation within International River basins: the importance of institutional capacity. Water Res Updat 125:31–40

World Bank (2008) Agriculture and rural development in MENA. http://siteresources.worldbank.org/INTMENAREGTOPAGRI/Resources/AGRICULTURE-ENG-2008AM.pdf. Accessed 20 Dec 2017

Yoffe S et al (2003) Conflict and cooperation over International freshwater resources: indicators of basins at risk. J Am Water Resour Assoc 39(5):1109–1126

Part II
Transboundary Hydro-Governance

Chapter 4
Understanding Hydro-Governance

What makes the desert beautiful, said the little prince, is that somewhere it hides a well.

Antoine de Saint-Exupery, Author

Abstract The purpose of this chapter is to examine and understand what is a governance of water, or hydro-governance, enabling to deal with all kind of waters, i.e. surface, groundwater and coastal water. Having proposed clear distinctions between the concepts of Management, Policy and Governance, often somewhat confusing, and analyzed some basic tools used to implement these concepts, such as, for instance consultation and public participation, this chapter introduces "hydro-governance" as a complex, multi-disciplinary and multi-scale socio-economic process for managing water related issues. This specificity is related to the nature of water governance, illustrated by its interaction with different scientific disciplines, many socio-economic and professional sectors at various scales and its strong environmental and climatic variability. Several definitions of hydro-governance have been introduced in the past, some of which being adopted by various international organisations, such as the OECD, UNESCO, UNDP or the EU. These are presented and compared, leading to proposing a realistic and effective hydro-governance. Also, the specificities of groundwater resources are considered, justifying the introduction of a specific groundwater-governance.

Finally to enable the use of hydro-governance as a tool in real complex water related conflicts, the largely theoretical approach adopted in this chapter is tempered by the presentation and analysis of major deficiencies, drawbacks and issues, with again some specific aspects concerning groundwater-governance.

Keywords Governance · Policy · Management · Water · Groundwater · Public participation · Scaling

J. Ganoulis, J. Fried, *Transboundary Hydro-Governance*,
https://doi.org/10.1007/978-3-319-78625-4_4

4.1 The Concept of Governance

4.1.1 Beyond Management and Policy

To understand the meaning of "governance" we should refer to its etymology, in which the term derives from the Greek verb "kybernao" that means "steering". For example, the steersman of a boat is in Greek a "kubernētēs" and "cybernetics", a term popularised by the mathematician Norbert Wiener in the 1940s, is related to the idea of governing. From steering a vehicle or a boat, the term "governance" took by Platon, one of the fathers of Greek philosophy, its metaphorical sense in political sciences, as *the art of ruling a society*.

Only recently, as in the 1990s, this term was disseminated by international institutions like the UN, the IMF and the World Bank, to a particular 'level', "type" or "field" of activities. We then talk about "local or global governance", "project governance", "economic governance", "environmental governance" or "electronic governance". In this book, we deal with "water governance" or "hydro-governance" and we are also referring to a particular "model of governance", such as "authoritarian, collaborative or participatory and multilevel governance".

In all these cases, the picture of steering a sailboat, as shown in Fig. 4.1 is very useful for understanding the meaning of "governance" and its difference from "management" and "policy". In fact, steering a sailboat is a complex process for achieving by successive approximations a particular goal, under different external adverse pressures. The goal is to arrive safely to a given destination against unfriendly winds, waves and marine currents by using different tools, like the sail, the steer and the electronic instruments of the boat. In this complicated travelling process (governance), different decisions are to be taken, e.g. how to undertake particular steps (management) following a given set of rules, such as navigation standards and laws (policy).

From the various concepts above, we may resume metaphorically the following definitions:

Governance is the political process for steering a society.
Management is the decision-making of particular measures for resolving societal issues at different levels.
Policy is the set of laws, regulations and customs for managing societal issues.

In a sense, governance contains and integrates the notions of management and policy.

As we will explain later in this book, *Good Governance* is the science-based, multi-level, interactive and integrated political process for efficiently managing complex societal problems and issues, taking into account a given policy. This process is iterative and "good" governance can improve policy.

Governance, Policy and Management are three interrelated concepts, which are often used indistinctly in a confusing manner. Further clarification is therefore needed for defining them in a more general way and, more specifically in the context of this book, when applying them to water governance.

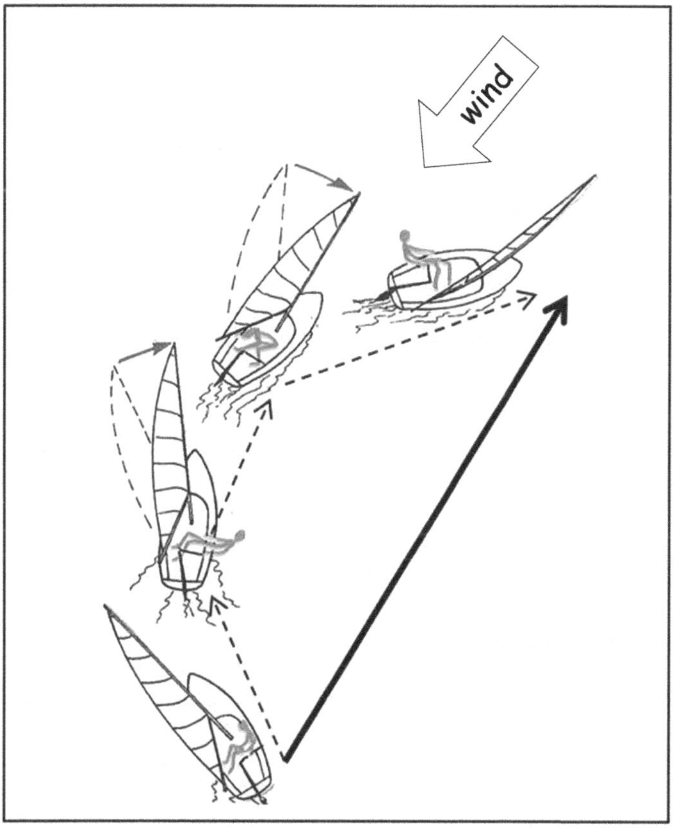

Fig. 4.1 Steering a sailboat by using different tools against the wind and other pressures

Through discussions with many colleagues, we have concluded the following more general definitions of policy, management and governance:

Policy refers to a vision of a society and its objectives; it sets the laws, regulations and customs in order to reach these objectives and fulfil the vision.

Management refers to the process of coordinating, planning and even controlling information, institutions, activities, functions, and efforts to achieve the objectives of a policy.

Governance is the hardest of the three concepts to pin down. It refers to the use of policy, management, public consultations and participation, and other tools and mechanisms to interact and direct people, organizations, and political units. It also refers to the interactions between those who are being governed and individuals, groups, NGOs, and institutions that are governing. At its core, governance is the *basic process* that encompasses, coordinates, and controls the use of policy, management, and other interactive tools and mechanisms that connect the governed and the governing stakeholders.

The analytical definitions given above constitute a basis for better understanding the meaning of these three concepts that we will use for introducing the notion of "hydro-governance", i.e. the governance of water resources use and its relationship to policy and management. Before analysing the "hydro-governance process", we will clarify different types of governance and stress the meaning of public consultation and participation, as an important tool to improve the implementation of governance, policy, and management.

4.1.2 Governance Systems

Although governance is not synonymous with government it is very much related to the mechanisms governments follow to take decisions for setting up or modifying a national policy framework. In this sense, governance is very much related to the country's political system and the decision-making processes.

In terms of different political systems, two of them are historically well known and are in place in the majority of contemporary states: (1) the *authoritarian* one and (2) the *democratic* political system. Both systems are applied with different connotations and different characteristics, depending on the historical and socio-economic characteristics of each country.

The authoritarian system is more or less common to many modern dictatorships and has similar characteristics with the ex-communist regime, which officially is still active in North Korea. A simplified view of an authoritarian model for "Water Supply Governance" is shown in Fig. 4.2. Although the decision-making processes

Fig. 4.2 Authoritarian model of water supply governance

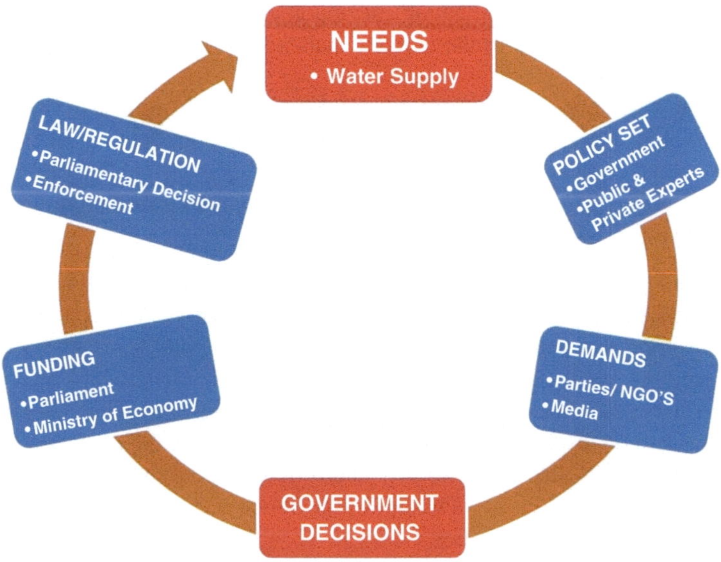

Fig. 4.3 Democratic/Parliamentary model of water supply governance

are much more complicated, this picture depicts the main steps taken by the main actors in order to take relevant decisions. The role of the government in this system is crucial and we can say that this is a *state-centric system* of governing.

As we can see in Fig. 4.2, the national policy is set by the Government, that in authoritarian systems is composed by one party. This Government elaborates a x-years policy plan that responds to demands from the party and the party's controlled media. After the necessary funding is released from the Political Bureau and the responsible Ministry, the corresponding law/regulation is entered into force.

The corresponding *classical democratic model* is shown schematically in Fig. 4.3 for the specific case of Water Supply Governance. Although in this case the succession of particular steps for developing the decision-making process is the same as in the authoritarian system shown in Fig. 4.2, the involvement of additional actors, such as different political parties, NGOs, private experts, media of different sensitivity and the exhaustive parliamentary debates make the difference between these two models of governance.

The results obtained by the democratic system are ideally monitored and evaluated, and are constantly subject to criticism from different social actors, such as the opposition parties, NGOs, civil societies, public or private associations and simple citizens. The degree of public consultation and public participation in the decision-making processes is one of the key characteristics of the democratic model of governance and the guaranty of its effectiveness.

There is also a need for improving and revitalising the old model of democracy in our new global economy and modern socio-economic conditions. Improving transparency in the decision-making processes, reducing the authoritarian behavior of the majority-based politicians, combating corruption and ensuring a fair distribution of economic growth are some of the modern challenges for improving the democratic model of governance.

Because water and environmental issues are intimately linked to a variety of socio-economic activities, such as agriculture, energy production and industry, "environmental" and especially "water governance" become challenging issues as traditional concepts of government are changed. Governments face different changes that arise at many levels:

- At the *international level*, national governments are losing power and the exclusive authority for protecting national interests, because of the economic globalisation, improved communications and increasing capital flows across national borders.
- At the *national level*, the private sector is increasingly involved for providing services and policy in areas traditionally under governmental control.
- At the *local level*, the public society and different community associations and networks replace traditional activities of the public sector and are weakening the influence of national authorities.

In the water sector, which is very much related to food and energy production, all these changes require a rethinking of social processes and the structure of economic activities, reflected in the development of new concepts of "water governance" and "groundwater governance".

4.1.3 Towards a Participatory Governance

One particular characteristic of "Good" Governance is more and more related to public participation, and the United Nations and various international institutions have reinforced the importance of public participation, offering guidelines, handbooks and other types of assistance. In many countries, laws and regulations regularly require public meetings and comments on government actions. Public participation is fundamental to maximize agreement, enhance transparency and decision-making, create ownership and facilitate the acceptance and enforcement of decisions and policies. It is also a mechanism for gaining a better or common understanding between the various stakeholders on the nature of a given problem and the desirability of specific outcomes. Stakeholder participation strengthens integration, thereby contributing to conflict prevention, and risk reduction—all highly important in large infrastructure development projects. Active citizenship is also a way to stimulate the interest of the political representatives, policy-makers and decision-makers in water governance and management, which is very important to guarantee a continuous political involvement, knowing that such an involvement is needed to obtain financial support on the long term. The various levels and ways of public participation are summarized in Fig. 4.4, from simple information to elaborate active involvement.

Fig. 4.4 Different ways of public interaction and participation

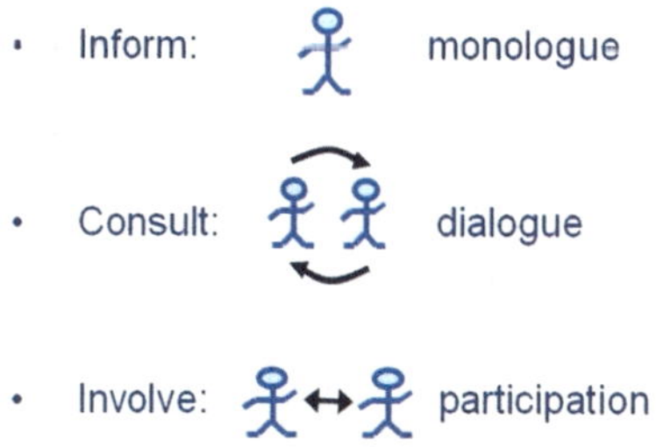

But, as emphasized by the US Environmental Protection Agency (US EPA), meaningful public participation requires much more than simply holding public meetings or hearings or collecting public comment. In order to assist those who must manage processes where public input is important to decision-making, such as government agencies among others, the US EPA has produced a Public Participation Guide, aiming at helping to recognize the difference between meaningful public participation and less valuable forms of public engagement, and, most important, to select and design public participation programs to best meet the needs of given projects and better identify concerned publics. The EPA Guide (Public Participation Guide… 2017) provides a definition of public participation, a rather useful description of its benefits especially in terms of decision-making, and an evaluation of the conditions to make it a successful process. To illustrate the kind of information this EPA Guide can provide, hereafter we briefly quote its introduction entitled "Introduction to Public Participation" (it is a simplified quotation and not word by word):

"Public participation can be any process that directly engages the public in decision-making and gives full consideration to public input in making that decision. Public participation is a process, not a single event. It consists of a series of activities and actions by a sponsor agency over the full lifespan of a project to both inform the public and obtain input from them. Public participation affords stakeholders (those that have an interest or stake in an issue, such as individuals, interest groups, communities) the opportunity to influence decisions that affect their lives. It results in better governance, especially enabling sponsor agencies to make decisions that reflect public interests and values and are better understood by the public. A successful public participation requires the appropriate level or degree of public participation for the decision at hand and the setting of public's expectations accordingly, meaning that agencies should promise what they are willing to commit to and deliver and not more". As shown in Fig. 4.5 for the specific case of Water Supply Governance, public consultation and participation is one of the basic elements of all stages enabling decision making, management and policy.

Quick and Bryson (2016) have analysed public participation from a more academic point of view than the EPA Guide, providing a rather complete basis for this significant component of governance.

As we consider their analysis to be a useful preparation to the application of public participation to transboundary situations, hereafter we briefly present their main results as simplified quotations of their article:

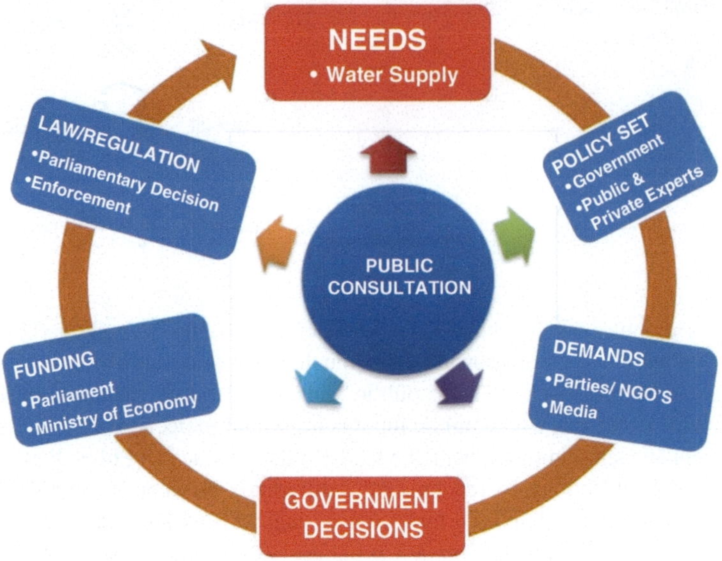

Fig. 4.5 Participatory model of consultative governance

Although by the early 2000s, public participation had become a routine and expected feature of public policy-making (Bingham et al. 2005), and its practice has become increasingly professionalized, the methods of participation vary greatly, so that there is great interest in how well it is designed and implemented (Sandfort and Quick 2013). Not only the dimension of participants' influence on decision outcomes should be considered, but also what kinds of stakeholders participate and how the participation is conducted. The dynamics of public engagement in governance are multi-dimensional and mutually constitutive. In fact, the implementation of participation in governance has important consequences for constituting the public. Sometimes participation processes are oriented to the general public, sometimes to the interested public and sometimes to smaller circles of representatives of key stakeholder groups. These variable definitions of the public and processes for representation introduce tensions in diversity and access to decision-making. Quick and Bryson (2016) propose four key elements for a successful public participation:

1. *Legitimacy*, expressed in terms of the adequacy of participation or representation, the technical or political workability of the decision outcomes and the procedural fairness of the process.
2. *Diversity and Inclusion*, as a key challenge is to ensure that an appropriate range of interests is engaged in the process, including under-represented and marginalized groups, and not underestimating the difficulty and the possible negative results of gathering groups with ethnic, racial, gender or socio-economic diversity.

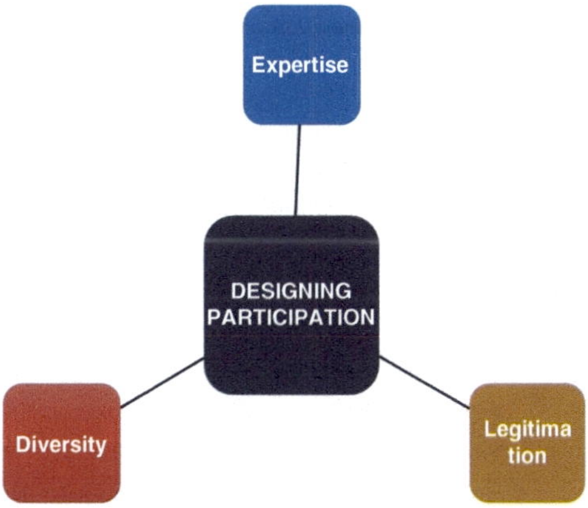

Fig. 4.6 Elements for designing a successful public participation process

3. *Expertise and Participation*, as decision-making may rely on expert knowledge based on science and a neutral approach while the concerned public does not necessarily have all dimensions and aspects of the discussed problem in mind, or may be influenced by its personal interests, although the public may also be closer to the real issue than the expert and bring a different approach.
4. *The Challenge of Designing Participation Processes*, as there is no formula for good participation.

The latter consideration is the base for building up a successful consultation, taking into account the three preceding elements (Fig. 4.6).

Quick and Bryson conclude by proposing two areas of research and thinking: first, how much participation is desirable and workable, as public participation is not easily accomplished and may not always be appropriate. And second, the implications of increasingly diffuse systems of governance for opportunities for public participation, such as the increased involvement of non-governmental entities in public services among others.

From these various comments on "Public Participation and Consultation for Decision-Making", we can say that numerous methods exist for public involvement ranging from compilation of a stakeholders' database for network interaction to public hearings. Participation should be organized in an open and transparent way and should involve all relevant groups. Residents, government representatives, the research community, farmers, industries, the private sector, women and minority groups all need to be fully involved in the development of river basin, lake and aquifer strategies, agreements and institutions. Of course, numerous challenges to public participation exist as well, for instance differing legislation and management and public participation systems—as well as priorities—in neighbouring countries. Frontiers frequently represent a "delimiter" not only of a linguistic but also a cultural and socio-economic nature, and the public can be insufficiently aware of how to take part in decision- making. In addition, mechanisms of public participation are not well developed in many countries and even less at the transboundary level.

Critically, public participation requires adequate financial resources to be effective. Yet despite the difficulties, transboundary public participation efforts can be successful: witness the Danube Convention, the Sardar Sarovar Project and the Regional Partnership for Prevention of Transboundary Degradation of the Kura-Aras River. (UN Water Thematic Paper 2008).

To summarize, we will answer two questions: (1) who is the public and (2) why is public participation so important?

1. The public consists in:
   - the '*general public*', the population living within a river basin
   - those *individuals* and *organisations* with a specific interest in water resources management, potential investors.
   - Those that can really contribute to delivering solutions: water *companies, wastewater treatment companies*
   - NGOs, (Non-Governmental-Organizations) such as those promoting environmental protection.
   - *sectoral stakeholder groups* within a river basin (e.g. different water users such as agriculture, industry etc.)
   - *advisory committees*, in local level
2. Why is public participation so important?
   - Public consultation strengthens democracy.
   - Helps determining priorities, enhances understanding of problems and challenges
   - The involvement of the public makes people share responsibility and provides locally held information and knowledge.
   - It also creates awareness and ownership of decisions.
   - Provides transparency of actions and accountability.
   - Public participation promotes consensus building
   - It improves cooperation and coordination of actions.

Clearly, public participation is an important component of governance, hence hydro-governance, but it has been noticed that the public does not necessarily understand the significance of its participation and, often, it must be stimulated by, for instance, financial incentives, projects with a clear explanation of objectives, or new perspectives, and comforted by specific education (Dhanya and Renoy 2017).

4.1.4 Definition of "Good" Governance

The UN's Food and Agriculture Organization (FAO) proposes a simple definition of policy, as "a set of decisions which are oriented towards a long-term purpose or to a particular problem". The simplicity and succinctness of FAO's definition captures the centrality of the vital term, "decision", which is turned towards action, as shown in Fig. 4.5 for the participatory model of governance. Actually, for many policy-making it is synonymous with decision-making in all sectors of society at any level, with a stake in governance, and not just for public-sectors institutions (Varady et al. 2013).

Once policies have been formulated, putting them in place requires instruments, tools, rules, protocols, and other procedures. These may include laws, sets of rights, registrations, permits, and regulations (especially regulations that allow those regulated to choose among alternative ways of complying); economic incentives and disincentives such as subsidies, taxes, tradable pollution permits, and pricing structures; and civil-society actions such as those that motivate voluntary actions or behavioral changes. The institutions that enable all these procedures and instruments to work and interact and their organization and linkages build up the "governance".

In the field of corporate governance, in new public management, and in the "good" governance movement associated with the World Bank, the term tends to signify the pursuit of fiscally responsible, efficient, and accountable organizations (Fig. 4.7). Best business practices are advocated as a means to effective governance.

As clearly expressed by Tropp (2007) and Rogers and Hall (2003), "governance is about the processes of making choices, decisions and trade-offs. Governance addresses linkages and processes between and within organisations and social groups involved in decision-making, both horizontally across sectors and between urban and rural areas, and vertically from local to international (Rogers and Hall 2003). Governance is not limited to "government" but also includes the private sector and civil society. The character of relationships (and the formal and informal rules and regulations guiding such relationships) between different social actors and organisations is an important feature of governance".

Fig. 4.7 "Good" Governance definition in corporate business

4.2 Hydro-Governance

4.2.1 A Multi-disciplinary Integrated Approach

"Hydro-governance" relates to the range of political, social, economic and administrative systems that are in place to develop and manage water resources and the delivery of water services at different levels of society. Or put more simply, "hydro-governance" is the set of systems that control decision-making regarding water resource development and management. Hence, "hydro-governance" is much more about the way in which decisions are made (i.e. how, by whom, and under what conditions decisions are made) than the decisions themselves.

To understand "Hydro-governance" it is necessary to remind the concept of water management as a multi-disciplinary integrated framework. As shown in Fig. 4.8, the *Integrated Water Resources Management* (IWRM) approach could be achieved by coordinating different sectors (e.g. drinking water supply, agriculture, industry, tourism, public health, ecosystems), different scientific and professional disciplines (e.g. engineering, geology, biology, chemistry, economy, social sciences), different institutions (e.g. public and private research centres and institutes, laboratories), different stakeholders (e.g. associations, NGOs, companies) and various impacts

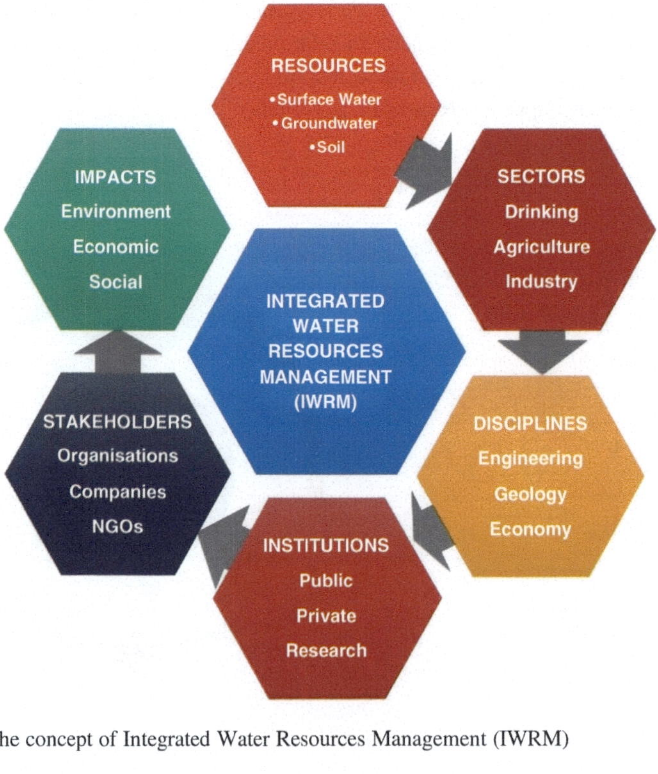

Fig. 4.8 The concept of Integrated Water Resources Management (IWRM)

(e.g. environmental, economic, social). There is no general rule about the optimum degree of integration and how to achieve it.

Hydro-governance covers the way allocative and regulatory politics are exercised in the management of water and other natural resources and broadly embraces the formal and informal institutions by which authority is exercised. The relatively new term for discussing this combination of formal and informal institutions is *distributed governance*.

There is a profoundly political element to hydro-governance and, as such, systems of hydro-governance usually reflect the political realities and choices at international, national, provincial and local levels: for instance, a socialist approach with all aspects of water being public, or a liberal approach with water managed privately, or in between with public-private partnership and a balance of the public and private sectors.

We propose to leave all options open and adopt a rather general definition due to the Global Water Partnership: Hydro-governance is the "range of political, social, economic and administrative systems that are in place to develop and manage water resources, and the delivery of water services, at different levels of society" (Rogers and Hall 2003). OECD has further explained that Hydro-governance encompasses administrative systems, formal institutions (including laws and policies) as well as informal institutions such as power relationships and practices. For the Stockholm International Water Institute (SIWI), hydro-governance refers to the political, social, economic and administrative systems in place that influence water use and management. Essentially, who gets what water, when and how, and who has the right to water and related services, and their benefits. SIWI further explains that Hydro-governance "determines the equity and efficiency in water resource and services allocation and distribution, and balances water use between socio-economic activities and ecosystems. Governing water includes the formulation, establishment and implementation of water policies, legislation and institutions, and clarification of the roles and responsibilities of government, civil society and the private sector in relation to water resources and services. The outcomes depend on how the stakeholders act in relation to the rules and roles that have been taken or assigned to them. The water sector is a part of broader social, political and economic developments and is thus also affected by decisions by actors outside of the water sector".

For the specific case of the European Union, the European Commission defined 'European governance' in 2001 as referring to the rules, processes and behavior that affect the way in which powers are exercised at European level, particularly about openness, participation, accountability, effectiveness and coherence. Concerning water, the 2012 EU Task 1 Governance Final Report (EU Commission 2012) has a more restricted approach than the broad definitions written above. It focuses on authorities and legal frameworks for water management—and thus at rules and processes, cited as a key element of European governance. However, it uses a different set of criteria to focus on issues in the water sector. Although this report does not focus on accountability, it is an important element in governance overall and for the Water Framework Directive. For instance, the World Bank describes accountability in the following terms: "Accountability ensures actions and decisions taken by public officials are subject to oversight to guarantee that government

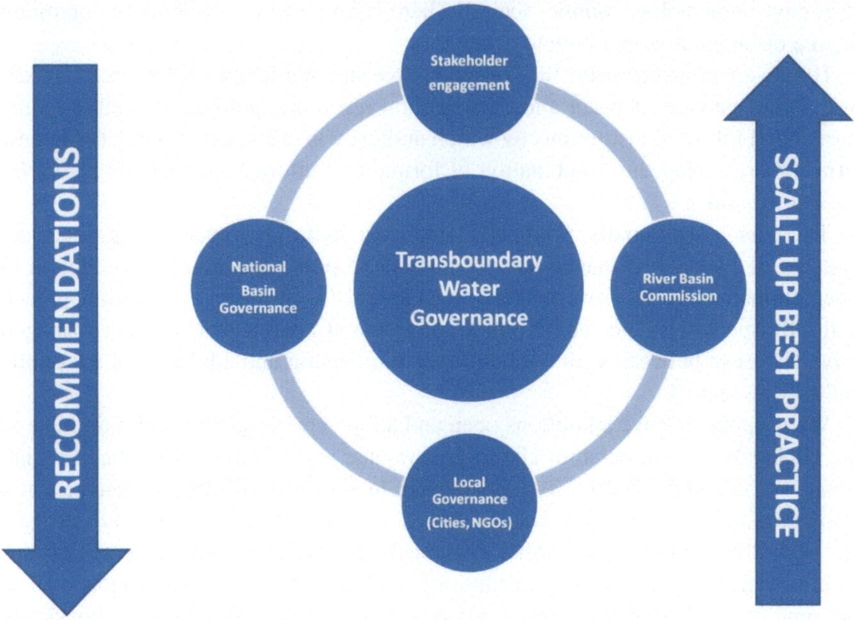

Fig. 4.9 Gaps in water-governance (OECD 2011)

initiatives meet their stated objectives and respond to the needs of the community they are meant to be benefiting". This approach further divides accountability into two stages: 'answerability', the obligation of government to provide information about decisions and actions; and "enforcement" when there are violations of this obligation.

The OECD has proposed an analytical framework and tool for policymakers to identify and bridge governance challenges called "OECD Multi-level Governance Framework: Mind the Gaps, Bridge the Gaps", which, according to OECD, applies to all countries "regardless of their institutional setting, water availability or degree of decentralisation". It is summarized in Fig. 4.9, where the gaps affecting the design and implementation of a water policy are identified and an attempt at an analytical framework is proposed, which has been used in a number of countries to review their hydro-governance arrangements. This yielded the conclusion that governance strongly depends on existing territorial specificities and that no "one-size-fits-all" solutions did exist.

4.2.2 *International, National, Regional and Local Levels*

How is water governed? At which level? Is hydro-governance a theoretical general concept or does it depend on its scale of application, local, regional, national, international or transboundary? "Scale" is a concept used by social scientists (especially geographers) to refer to the relationship between geographic (environmental) units and political (human) units of all sizes (Norman et al. 2012). Here we use scale in its classic cartographic meaning, to differentiate between local, regional, national and international governance and management.

Moore (2013), whom we quote hereafter, proposes some answers:

> State and sub-state governments have the official authority to govern water in most watersheds throughout the world, although they may choose to delegate, rescind, or even neglect that authority. However, the global water governance literature indicates that the complexity of water governance relates to the fact that, at the organisational level, the authority for steering the water agenda is lacking. At the global scale, a single regime for governing water does not exist (Conca 2005). However, in the past two decades, institutions in the global domain have begun to proliferate, and include agencies such as the World Water Council, UN-Water, and the Global Water Partnership. But other organisations at the international level, such as the World Bank, the FAO, and the UNDP, also have an enormous impact on water development projects and the governing conditions that surround these projects. Yet, these institutions are not by any means intended to be the lead agency for global water governance, nor are they solely focused on a water mandate. Moreover, other regional and national agencies have been heavily involved in shaping water governance for many countries and for many river basins. For example, a policy group, known as the Stockholm International Water Institute (SIWI), houses the Water Governance Facility—a joint initiative between the UNDP and the Swedish International Development Agency (SIDA). SIWI and the Water Governance Facility articulate their goals as providing support to developing countries for water governance reform through the provision of technical and policy support, advocating improved water resources management, and participating in global and regional water monitoring (Water Governance Facility 2011). Furthermore, SIWI has formalised many other partnerships, including with the Global Water Partnership, UN-Water, the International Water Association, and the Water Integrity Network hosted by Transparency International. SIWI has become involved in many of the global-level activities for water governance, including hosting the annual World Water Week, being an active contributor to the agenda of the World Water Forum and initiating new dialogues and pilot projects aimed at reducing corruption in the water sector, a topic historically not considered in water policy debates. Yet, limited analysis is available that acknowledges SIWI as a stakeholder or authority figure in water governance, and little, if any, work has been completed to track its influence in water governance. This point is not intended to criticise SIWI's efforts; rather, it is simply one example that can be used to highlight the lack of clarity over which organisations or individuals at the global level are understood to hold authority, power, or influence in water governance. Without knowing who contributes to the governance of water, addressing its challenges becomes nearly impossible. However, while this lack of clarity about who has the authority to govern is a concern at the global scale, how this affects relationships amongst those responsible for water governance within individual watersheds at the local scale is particularly problematic.

Therefore, the problems of scale must be clarified to understand how and at which level an efficient water governance can be defined and implemented, and how the various levels are linked and coordinated. As shown in Fig. 4.10, different authorities and administrations are responsible for water management problems at different levels, such as the River Basin, the local administrative level, the regional level and the central national level (Government and Ministries). Although these authorities take different forms from country to country and their legal status (public, private or mixed) depends on the particular country's regime (federation, one central or regional governments), *water governance is the art of coordinating them in order*

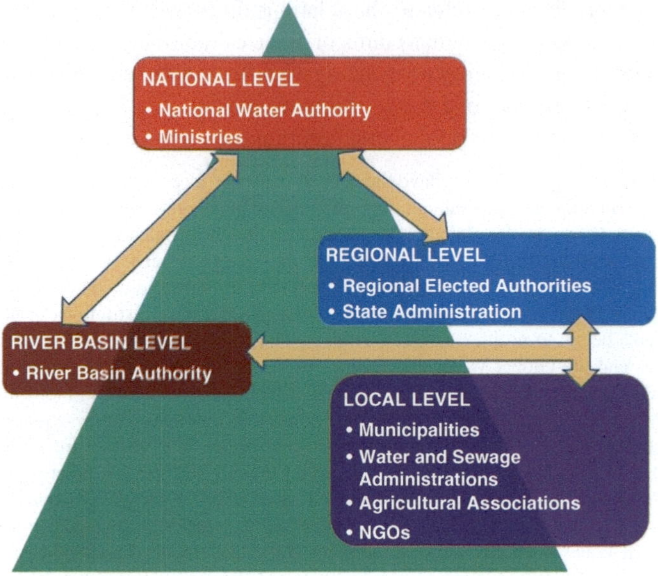

Fig. 4.10 Hydro-governance at national or state scale

to ensure "water security" at all levels and protect population from floods and droughts.

This is a real challenge if we consider the necessary adaptation against climate change and the multi-disciplinarity of the water management process (see Fig. 4.8) at every level shown in Fig. 4.10. To obtain water security and water saving from over-consumption, especially in agriculture, a unified, integrated and decentralised water governance should be set under the state own responsibility at all levels.

A workshop organized by UNESCO in 2011 tried to answer such questions by examining how water governance is evolving at different levels of governance and what are the implications for principles of good governance such as equity, legitimacy, efficiency, transparency and accountability. One of the presented papers (Toonen 2011) discussed the concept of *Multi Level Governance* (MLG) to illustrate the changing role of national governments in water governance defined as "being about establishing the appropriate means and setting limits and constraints within which operational action in terms of water management can take place, water management being about achieving goals, preferably in a functionally and socially responsive and efficient manner...".

We consider the conclusion of this paper as quite instructive as it rather well summarizes what is MLG in a useful way:

> Modern water management in MLG is fundamentally bottom up, founded on a strong societal self-organization. Human and social capital are as important as money, and form the key to good water governance. This value can be seen in poorer countries with less material and economic resources than the Netherlands, and therefore more dependent on

creativity and innovation. Decentralized systems can in fact be very structured, well-organized and effective. The interests of all relevant actors should be considered. Solutions to improve the system should be based on consensus building and negotiation. Many systems come from (or are still in) a transition process from the centralized tradition of the ('unified/federal') nation-state. The future of international water management lies with stewardship, smart water grids, stakeholder involvement, ('federalized') catchment area organizations, water- and user—boards, large infrastructure management (also by private firms), public-private partnerships, well-regulated markets and a strong involvement of financial institutions ranging from micro finance for small-scale social start-ups to large-scale strategic asset management for the maintenance of these large infrastructures.

Water safety and water quality are constantly under threat of 'the tragedy of the commons' as well the daily neglect of maintenance by 'politics as usual'.

It takes a strong, solution driven, internationally oriented, professionally qualified, bureaucratically competent and politically proficient national government organization to deal with it.

4.2.3 Scaling Problems and Trends

For some years now, it has been recognized that governance in the domain of water is related to the institutional levels of decision-making, local, regional, international, and the relationships between these levels expressed as relationships between geographic units and political units of all sizes. The sizes of these units, their relationships and the relationships between their sizes constitute the scale problems, which have been recognized by several authors as a quite important issue for social scientists engaged with water issues (Norman et al. 2012a, b). E. Norman, K. Bakker and C. Cook state that scale is an important issue because "significant scalar reforms have taken place in the past few decades. Although varied, these reforms often entail devolution to lower scales of management, increased citizen participation, new decision-making processes, and new types of community organisations (e.g. watershed committees)".

Another scale aspect of hydro-governance is the concept of *Global Water Governance*, (GWG) defined and presented by the Pacific Institute Report (Cooley et al. 2013): "Throughout the twentieth century, water governance efforts emphasized the local and/or regional scales, in part because water challenges were largely perceived as local issues. However, there is growing recognition that the scope and complexity of water-related challenges extend beyond national and regional boundaries and therefore cannot be adequately addressed solely by national or regional policies. In particular, widespread water scarcity and lack of access to water supply and sanitation threaten socioeconomic development and national security for countries around the world. Additionally, people around the world share and exchange water directly and indirectly through natural hydrological units and systems and through global trade, i.e. virtual water. Furthermore, climate change and the growing presence of multinational companies within the water sector play a role in globalizing water issues. Despite the need, discussions about GWG have been limited. Water governance studies that have taken a broader perspective have largely focused on transboundary water resources (Schmeier 2010). GWG has also been discussed within other more prominent global governance challenges (notably climate change and energy) and within discussions of global development objectives (Florini and Sovacool 2011; Tropp 2007). However, there has been little-to-no discussion about GWG that looks more holistically at global water challenges and the structures and approaches needed to meet these challenges. One of the few definitions of GWG comes from a 2008 study (Pahl-Wostl et al. 2008) that defines it as "the development and implementation of norms, principles, rules, incentives, informative tools, and infrastructure to promote a change in the behavior of actors at the global level in water governance". Thus, GWG focuses on the processes of international cooperation and multilateralism. It comprises formal and informal instruments—including global governmental and non-governmental organizations, regimes, actors, frameworks, and agreements—created to balance interests and meet global water challenges that span national and regional boundaries. GWG informs how challenges are being tackled (or not) at the regional and international levels among various players (from governmental bodies to civil society organizations) and suggests opportunities and barriers to meeting global objectives. GWG also facilitates interaction and dialogue among key players to inform the development of solutions to problems at local, national, and regional levels to ease global pressures.

From a transboundary perspective, the scaling problems do affect hydro-governance issues and must be accounted for, as illustrated by Jacobs (2012), who describes the multi-levelled development of institutional and legislative instruments for the governance in the Orange-Senqu and the Southern African Development Community (SADC), from the International, Regional, Basin and National Levels, and the interactions between levels.

The *international level of scale* is influential in so far as it provides a non-binding, guiding framework at the global level for the management of transboundary water resources. As shown in Fig. 4.11 starting from the national water governance, the international scale is about the "transboundary hydro-governance" that will be analysed in Chap. 5 that follows and whose groundwater aspects can be found in (Conti 2014).

The global picture of transboundary water governance may indicate which countries have committed themselves to comply with the normative principles of transboundary co-operation, such as equitable utilisation, prior notification and the no harm doctrine. It also promotes the establishment of ad hoc watercourse agreements to be adopted for other regional watercourses and acts as a guideline for specific bilateral and multilateral regional agreements relating to the use, management and preservation of transboundary water resources.

The regional legal framework reflects the international context in terms of the adoption of the global principles of equitable utilisation, no harm and prior notification found in several international instruments including the UN Convention as well as the Helsinki Rules. The 2000 Revised Protocol on Shared Watercourses (SADC 2000) is a valuable legal instrument, which has given rise to several regional

Fig. 4.11 National and international scales of transboundary hydro-governance

plans and strategies including the Regional Water Policy (RWP), the Regional Water Strategy (RWS), as well as the Regional Strategic Action Plan (RSAP I and II) that together have provided the roadmap for the implementation of the SADC Revised Protocol.

Institutional arrangements at *the basin level* have evolved over time and reflect the changing political, social and economic transformations that have occurred in the region (Kistin and Ashton 2008). These co-operative arrangements were established in a time when public participation was not routinely performed and environmental accountability was unheard of. The four riparian states (Lesotho, South Africa, Namibia, Botswana) have established several bilateral agreements and one multilateral basin-wide treaty of noteworthy importance (Kistin and Ashton 2008).

While the *national institutional and legal frameworks* vary between the four basin states, one shared factor is the transitional nature of these national legislative frameworks as reflected in the promulgation of new acts, the revision of old ones and the development of water policy and strategies. With few exceptions, these new pieces of legislation generally comply with the principles of the Revised SADC Protocol and directly (in the cases of South Africa and Namibia) and indirectly (in the cases of Botswana and Lesotho) with the UN Convention. The new or revised national water acts also make provision for international rivers and meeting international obligations, elements not previously found in the older national water laws.

4.3 Actual and Effective Hydro-Governance

4.3.1 Joint Surface and Groundwater Hydro-Governance

In terms of quantity, groundwater resources are globally by far the most important source of freshwater. Around 97% of the word's available unfrozen freshwater is stored underground, which means that groundwater is the most important potential source of freshwater for human use. Worldwide, 2.5 billion people depend solely on groundwater for their daily needs and groundwater supplies nearly half of all drinking water in the world (UN WWDR 2003). Although, within the Integrated Water Resources Management (IWRM) framework, groundwater should be considered together with surface water in a unified way, the specificity and complexity of groundwater resources may justify "groundwater governance" as a distinctive concept.

According to an article by Wijnen et al. (2012), IWRM is the basis of integrated surface water and groundwater joint governance that can allocate water in accordance with the institution's policy goals. It is supported by a legal framework comprised of laws and regulations, including possible existing rights, and an economic policy proposing incentives for good management, subsidies, taxes, cost-recovery and benefit-sharing methods, among others. With a framework which facilitates groundwater management at the lowest appropriate administrative level and supports local management, joint governance will be driven by knowledge and science, based on data collection, processing and interpretation by way of monitoring networks, stakeholders' observations, and modelling. Information and communication will provide the links between the scientists, the managers and the decision-makers and, of course, contribute to the positive involvement of the public and the various stakeholders in the governance process.

Although groundwater represents a major water resource in the world, in most countries it has not been studied and accounted for in the same way as for surface water. The reasons are multiple, and among others:

- Its invisibility which makes it much less familiar than surface water
- The ultimate linkage between groundwater management and land use
- Difficulties and costs of monitoring and financing groundwater resources studies
- Myths of its high quality and low vulnerability, as it is often considered as protected by the soils above it
- Weakness of institutions responsible for groundwater investigations
- Inadequacy of infrastructure and tools needed for integrated groundwater resources management
- Lack of models for international groundwater management commissions in transboundary aquifers, which is not the case for surface waters (e.g. the International River Commissions).

For instance, the European Union only started its groundwater resources survey in 1977 after the 1976 drought. At the same time, it started preparing the directive on groundwater protection against pollution (a EU "directive" becomes a law when transposed to Member States' legislation), which was adopted only in 1980 and updated in 2006 (2006/118/EC Groundwater Directive to WFD 2006).

The management of groundwater resources is not an easy task and when it is complicated by the presence of political and/or administrative boundaries, mainly in countries with federal structure, it can easily become a cause of national or even international conflicts. Regional environmental and human security and peace are at stake, so people should learn how to cooperate for the best management of their common water resources, in application of the UN sustainable development principles. Essential characteristics of groundwater are endangered due to the deterioration of water quantity and quality and vital functions of groundwater reservoirs are threatened by pollution and overexploitation. Many examples can be given from around the world, as the aquifer depletion in the plain of Thessalia, Central Greece due to agricultural water overuse and the salinisation of coastal aquifers in the Mediterranean due to over pumping (Ganoulis 2016). A significant cause of the deterioration of groundwater quality is increasing pollution from diffuse sources, like agricultural activities and groundwater nitrification due to over-fertilisation in agricultural areas.

The protection of groundwater resources may be based on different methodologies involving either simple empirical or sophisticated methods. Various traditional strategies for groundwater protection range from the construction of groundwater vulnerability maps and the definition of protection perimeters around pumping wells, to the use of sophisticated optimisation multi-criterion decision-making techniques under risk conditions. A very characteristic example is the definition of adequate waste disposal sites in relation to the risk of groundwater contamination. In many cases, problems and issues formulated as above might be resolved by technological or structural solutions. For example, it may be possible to construct bi- or multinational artificial recharge facilities in order to ensure adequate groundwater quantity for a region. Likewise, it is possible to develop water storage projects which serve entire watersheds and provide flood control, water supply, hydropower and groundwater recharge.

However, for such solutions to proceed, an appropriate cooperative framework must be in place. This includes state institutions, national and international legislation, economics and political involvement together with the participation of stakeholders and interest groups. This complicated framework is what is usually meant by the term "governance".

4.3.2 Towards an Effective Hydro-Governance

Surface water and groundwater should be jointly managed, and, in Sect. 4.3.1 we have discussed joint surface water and groundwater governance but stressed the originality of groundwater, especially in terms of physical characteristics, availability, protection and perception by both the policy-makers and the public. Besides, experience shows it is more complicated to govern and manage groundwater than surface water. Therefore, there could be a need to define and use a specific governance for groundwater for a better efficiency.

For instance, it is always difficult to deal with something you cannot measure and, because groundwater is by essence invisible, it is rather hard to measure, hence to manage. This invisibility and the local and individual character of groundwater, still often owned by the owner of the land above, make groundwater users feel free from cooperation with their neighbors within a governance framework. Although it is a common pool resource, groundwater is easily appropriated just by drilling wells and capturing it without considering the neighbors, creating de facto rights under the land you own, which becomes difficult to deal with when a general legislation within a policy and governance project is adopted (Wijnen et al. 2012).

Therefore, governance today should take account of the reality that in many locations once groundwater rights have been asserted ahead of any governance systems that might have contained them, it is incredibly difficult to recover control. This is especially true in countries where all the incentives are in favor of development and abstraction, particularly where agricultural policy coincides with farmers' own motives to produce ever more, as the example of the rather new (2014) Sustainable Groundwater Management Act (SGMA) of the State of California shows with the challenges of its implementation. These external incentives are compounded by the powerful incentives inherent to the resource itself that lead farmers to prefer groundwater to all other water sources. Governance frameworks have proved very frail in the past to resist such powerful motives. Traditional and local governance developed to manage springs or oases have rarely been able to adapt to the new tubewell technology. Very few governments have been able to align agricultural policy with good water resources management, and even fewer have been able to recover control over groundwater once that control has been lost.

A Specific Governance for Groundwater

The human use of groundwater and aquifers is presently governed by a mix of social customs, codes, laws and regulations. Achieving a state of 'good' groundwater governance is a relatively complex concept and therefore there is considerable variation among the definitions of groundwater governance presented by different authors. Hereafter, we present some of these definitions, from the very simple to the more complete.

Definition 1: Groundwater governance comprises the enabling framework and guiding principles for collective management of groundwater for sustainability, equity and efficiency.

Definition 2: Groundwater governance comprises the enabling framework and guiding principles for responsible collective action to ensure control, protection and socially-sustainable utilization of groundwater resources for the benefit of humankind and dependent ecosystems.

Definition 3: Groundwater governance is the process by which groundwater resources are managed through the application of responsibility, participation, information availability, transparency, custom, and rule of law. It is the art of coordinating administrative actions and decision making between and among different jurisdictional levels—one of which may be global.

Definition 3, due to (Saunier and Meganck 2007), shows the difference between governance and management.

According to the UNESCO-GEF initiative "Groundwater Governance: a Global Framework for Action-A shared Global Vision for 2030", there exist four pillars on which Groundwater Governance should rely, as follows:

1. the actors, who formulate policies and strategies and are responsible for their implementation,
2. the legal and institutional framework, which rules how the different stakeholders interact,
3. information, knowledge and science, which drive the actors' decisions,
4. policies and plans, which define why activities are needed and when they should be undertaken or completed.

A fourth definition, highlighting these four pillars, has been proposed:

Definition 4: Groundwater governance comprises the enabling framework that establishes who formulates policies and strategies and is responsible for their execution (the actors) and how different stakeholders interact (the legal and institutional framework). Decisions made by the actors regarding what to do in pursuit of societal goals are driven by information, knowledge and science, and result in policies and plans which define why activities are needed and when they should be undertaken or completed. This framework determines the management of groundwater resources and the use of aquifers.

These definitions complete each other and a kind of practical general definition has been proposed by Varady et al. (2013), who defined groundwater governance as "*the process by which groundwater is managed through the application of responsibility, participation, information availability, transparency, custom, and rule of law. It is the art of coordinating administrative actions and decision making between and among different jurisdictional levels-–one of which may be global.*"

Implementation of Groundwater Governance

The practical implementation of groundwater governance involves identifying goals and policies, providing institutions and procedures, assigning dedicated personnel and financial resources, promoting stakeholder participation. The UNESCO-GEF initiative "Groundwater Governance: a Global Framework for Action-A shared Global Vision for 2030" has defined five basic principles that enable to practically implement groundwater governance, their fulfilment being more an aim for the water managers:

The first principle is that groundwater should not be managed in isolation, but conjunctively as appropriate with other water sources to improve water security and assure ecosystem health. Groundwater can often play the vital role of strategic reserve to cover variations in surface-water availability, and is both recharged by and discharging to surface-water bodies.

The second principle is that groundwater quality and quantity should be jointly managed, implying that groundwater management be harmonized with land management: land-use greatly influences how groundwater is recharged; therefore, groundwater recharge zones urgently need protection from pollution and degradation.

The third principle is that effective groundwater governance requires co-governance of subsurface space—a concept, which is still in its infancy, but needs to be developed.

The fourth principle concerns the need for 'vertical integration' between national and local levels in the elaboration and implementation of groundwater management and protection plans.

The fifth principle is coordination with the macro-policies of other sectors—such as agriculture, energy, health, urban and industrial development and the environment. In many cases policy action in these sectors holds the key to groundwater resource sustainability.

We wish to add that, compared to surface water, there are few legal and institutional tools designed to specifically manage groundwater resources and those that do exist are generally at the sub-national level and perhaps the aquifer or sub-aquifer scale (Linton and Brooks 2011; Feitelson 2003).

4.3.3 Main Deficiencies and Failures Worldwide

The purpose of Chap. 4 is to introduce the concept of hydro-governance and prepare the water managers for the application of this concept to transboundary waters. We have a rather complete and positive image of the concept in both theory and practice, but to be able to use it in practice, we must know how it has been applied until now, which means explore its deficiencies and failures.

To have a brief but significant survey of these deficiencies and failures, we have used two sources of information for water in general, the iied report "Water governance literature assessment" by Batchelor (2006), and the Pacific Institute report "Global Water Governance in the twenty-first Century" by Cooley et al. (2013). The Pacific Institute Report is rather extensive and its comments well organized and instructive; we have tried to extract the elements that correspond to our subject of governance, oriented towards transboundary situations. For groundwater, we have used the UNESCO-GEF Initiative Global Diagnostic (2015). Hereafter, we summarize the most important deficiencies given by these authors, essentially quoting the Pacific Institute Report:

- Intergovernmental Organizations lack clear leadership and coordination: for instance, UN Agencies have different agendas, sometimes overlapping, while bilateral funding agencies mainly care for their own priorities, and usually do not coordinate their efforts.

- The role of non-governmental processes is unclear: a broad range of new actors has appeared, coming from the private, non-governmental, academic and media sectors. They bring new ideas but their legitimacy, accountability and relationships with existing public governance structures are often unclear and may raise concerns about their actions with respect to government-led efforts.
- The lack of transparency leading to a lack of accountability can result in ineffective management strategies and investments, and stimulate corruption, especially related to the complex system of agencies responsible for its management and delivery; the growing presence of private actors and informal providers that operate in legal grey zones (where the actors are the de facto water service providers allowed to operate by governments but who may not have official license); and the large sums of money required for infrastructure investments.
- Political, legal and financial constraints play a role in the difficulties to agree on and to implement transboundary waters treaties to manage international river basins and aquifers. Besides, the lack of flexibility of interbasin agreements due to the absence of mechanisms to address changing social, economic or climate conditions considerably increase the difficulties to cope with inevitable changing climatic conditions.
- Funding in the water sector is considered as inadequate, despite various commitments of international institutions, like UNESCO for instance, and individual countries. Projects are launched but their follow-up is often not assured, because of money problems.
- Finally, education and training, especially in non-technological domains such as policy and governance, are not considered to be sufficiently important in various schools and universities to establish permanent courses and the corresponding needed funding is not easy to obtain from donors more used to the scientific and technological dimensions of water management. Besides, even these scientific and technological dimensions do not receive the attention and the funding they deserve, except when they are directly useful to a given local project. And, of course, we find similar difficulties with data collection and processing, as many regions of the world are still lacking basic water data and collection and, of course, correct maps. In that sense, groundwater is still a rather neglected resource compared to surface water and the main deficiencies and gaps in groundwater governance deserve a special mention.

Groundwater Governance Deficiencies

As already stated previously, groundwater governance has not been much on the agenda of decision makers. Despite the magnitude of its challenges and problems, as well as its importance to the livelihoods of a large proportion of the world's population, groundwater has not been subject to adequate policy and management attention, particularly compared to surface water. Groundwater has failed to feature prominently in water policy dialogues at local, national or global levels and, as a result, its governance has not kept pace with increasing demands and technological advances. It became evident that in most countries the ability to manage groundwater has not kept up with the increased dependence on this resource and with steadily growing pressures.

Almost everywhere groundwater governance is insufficient, with the following key deficiencies that have been identified in the UNESCO-GEF Initiative Global Diagnostic:

- Inadequate Leadership from Government Agencies
- Limited awareness of Long-Term Groundwater Risks
- No Measurement of Groundwater Resource Status
- Non-Performing Legal Systems on Groundwater
- Insufficient Stakeholder Engagement in Groundwater Management
- Limited Integration of Groundwater in Related National Policies

Hence there is an urgent need to strengthen groundwater governance—to address the current challenges and future significance of groundwater and to make use of opportunities for protecting and augmenting the groundwater resources. Because groundwater is essentially a local resource, effective governance arrangements need to extend down to the local level, but should also be linked to basin-level, national-level or even transboundary level, as appropriate. Good groundwater governance recognizes the value of aquifer systems, and aims at achieving the sustainable provision of freshwater and preventing the degradation of aquifer systems.

Yet the introduction of a specific groundwater governance should not hide the fact that groundwater quantity and quality strongly depend on the conditions of its recharge, particularly the state of degradation of the surface water systems above it. Shallow aquifers are especially concerned. Therefore, groundwater governance should serve as a platform for the implementation of *longer-term integrated water resources management (IWRM)*, including the sharing of groundwater resources among competing users (including the environment) equitably and transparently, and in conjunction with surface waters.

Groundwater Governance in the World: Current Status and Gaps

The current status of groundwater governance around the world cannot be easily summarized, although much information on groundwater governance has been collected by the UNESCO-GEF Groundwater Governance Initiative, essentially through five regional consultations and their various written contributions. But, because of the enormous diversity of conditions around the world, this project did not yet provide a systematic and reliable description of groundwater governance variation between regions and countries. Further, the information was often linked to a small area or to a special case, and sometimes open to interpretation. Nevertheless, two common lessons have emerged:

- *The state of groundwater governance depends very much on the groundwater management stage and on the economic conditions in the country concerned.* For this reason, groundwater governance principles are generally more embedded in relatively wealthy industrialised countries than, for example, in most Sub-Saharan African countries. Some of the greatest governance challenges are likely to be felt in rapidly developing countries in which governments can struggle to keep up with the rapid pace of change.

- *The focus of groundwater governance varies with the local needs and conditions.* Hence the focus on groundwater resource allocation in arid and semi-arid countries, on groundwater development for domestic water-supply in Sub-Saharan Africa, and on safeguarding groundwater irrigation for many millions of small farmers in South Asia. In Europe and North America increasing priority is being given to groundwater pollution control and aquifer and ecosystem protection, and some attention to the interaction with land-use and energy policies.

There are large governance gaps almost everywhere but the focus for future strengthening of governance varies according to development priorities and current risks. There is often a large gap between the present state and focus of groundwater governance and that which is actually needed and this is seen between and within the regions. Where development to meet basic water supply and sanitation requirements remains a major priority, much of the shorter to medium term governance focus needs to be on issues like financing, contracting and supervising such activities to ensure value for money, sustainability of supply sources and equity of service provision. There are also important governance requirements related to transparency and accountability, and the avoidance of corruption in the provision of contracted services. Many governments struggle to provide the institutional and technical capacity to meet these needs.

Overall, the state of groundwater governance is still rather poor in many countries and the strengthening of groundwater governance is a work-in-progress throughout the world.

There are many obstacles to adopting groundwater governance frameworks, particularly at the international level. Groundwater is a common-pool resource and is often utilized at an individual scale regardless of overall impacts to an aquifer. This approach creates a 'tragedy of the commons' situation, wherein the resource is utilized only regarding current benefits to the individual user. Meanwhile the likelihood for future, detrimental impacts to the resource and the users, as a collective, is ignored. The fact that most groundwater is accessible without many costs or requirements for monitoring exacerbates these issues. Negative impacts to the resource remain unseen and become evident only when adverse impacts to human and ecological health occur due to contamination and/or over pumping. In the absence of good groundwater governance and in the face of threatened transboundary aquifers, some states have experienced conflict and others have been motivated to seek out cooperative mechanisms for management, such as treaties.

References

2006/118/EC Groundwater Directive to WFD (2006)

Batchelor C (2006) Water governance literature assessment, 2nd edn. International Institute for Environment and Development, London

Bingham LB, Nabatchi T, O'Leary R (2005) The new governance: practices and processes for stakeholder and citizen participation in the work of government. Public Adm Rev 65 (5):547–558
Conca K (2005) Governing water-contentious transnational politics and global institution building, Global Environmental Accord. MIT Press, Boston. ISBN: 9780262033398
Conti K (2014) Factors enabling transboundary aquifer cooperation-a global analysis. IGRAC, Delft
Cooley H, Ajami N, Mai-Lan Ha, Srinivasan V, Morrison J, Donnelly K, Christian-Smith J (2013) Global water governance in the twenty-first century. http://www.pacinst.org/publication/global-water-governance-in-the-21st-century/. Accessed 19 Dec 2017
Dhanya V, Renoy G (2017) Water governance and public participation: what matters? Curr Sci 112 (4):686–687. https://doi.org/10.18520/cs%2Fv112%2Fi04%2F686-687
EPA Public Participation Guide: Introduction to Public Participation (2017) https://www.epa.gov/sites/production/files/2014-05/documents/ppg_english_full-2.pdf. Accessed 19 Dec 2017
EU Commission WISE (2012) Comparative studies of pressures and measures in the major River Basin Management Plans, Task 1 Governance, Final Report 28 Nov 2012. http://ec.europa.eu/environment/archives/water/implrep2007/pdf/Governance-Pressures%20and%20measures.pdf. Accessed 19 Dec 2017
Feitelson E (2003) When and how would shared aquifers be managed? Water Int 28:145–153
Florini A, Sovacool B (2011) Bridging the gaps in global energy governance. Glob Gov 17 (1):57–74. http://ink.library.smu.edu.sg/soss_research/2011. Accessed 19 Dec 2017
Ganoulis J (2016) Transboundary aquifer resources in South Eastern Europe (SEE): a regional case study. In: Fried J, Ganoulis J (eds) Transboundary groundwater resources-sustainable management and conflict resolution. Lambert Academic Publishing, Saarbrücken, pp 217–228
Jacobs IM (2012) A community in the orange: the development of a multi-level water governance framework in the Orange-Senqu River basin in Southern Africa. Int Environ Agreements 12:187–210
Kistin EJ, Ashton PJ (2008) Adapting to change in transboundary rivers: an analysis of treaty flexibility on the Orange-Senqu River Basin. Int J Water Resour Dev 24(3):385–400. Reflections on water management in South Africa. http://dx.doi.org/10.1080/07900620802127325
Linton J, Brooks DB (2011) Governance of transboundary aquifers: new challenges and new opportunities. Water Int 36(5):606–618. Strengthening cooperation on transboundary groundwater resources (Published online: 2 Sept 2011). http://dx.doi.org/10.1080/02508060.2011.599312. Accessed 19 Dec 2017
Moore ML (2013) Perspectives of complexity in water governance: local experiences of global trends. Water Alternat 6(3):487–505
Norman ES, Bakker K, Cook C (2012a) Introduction to the themed section: water governance and the politics of scale. Water Alternat 5(1):52–61
Norman ES, Bakker K, Cook C (2012b) Water governance and the politics of scale: how thinking critically about scale can help create better water governance. GWF Discussion Paper 1226. Global Water Forum, Canberra, ACT. http://www.globalwaterforum.org/2012/07/16/water-governance-and-the-politics-of-scale-how-thinking-critically-about-scale-can-help-create-better-water-governance/. Accessed 19 Dec 2017
Norman ES, Cook C, Bakker K (eds) (2012) Water governance and the politics of scale. http://www.water-alternatives.org/index.php/tp1-2/1881-vol5. Accessed 19 Dec 2017
OECD (2011) Water governance in OECD countries: a multi-level approach. OECD Publishing, Paris. https://www.oecd.org/governance/regional-policy/48885867.pdf. Accessed 19 Dec 2017
Pahl-Wostl C, Gupta J, Petry D (2008) Governance and the global water system: a theoretical exploration. Glob Gov 14:419–436
Quick KS, Bryson JM (2016) Theories of public participation (chapter 12 public participation). In: Torbing J, Ansell C (eds) Handbook of theories of governance. Edward Elgar Press, Cheltenham. https://www.researchgate.net/publication/282733927_Theories_of_public_participation Accessed 19 Dec 2017

Rogers P, Hall A (2003) Effective water governance. TEC Report No. 7. Global Water Partnership, Stockholm

SADC South African Development Community (2000) Revised protocol on shared watercourses. SADC, Gaborone. https://books.google.com/books?id=7gQmAQAAIAAJ. Accessed 19 Dec 2017

Sandfort J, Quick KS (2013) Deliberative technologies: learning from the art of hosting. Public Management Research Association Conference, Madison, WI

Saunier RE, Meganck RA (2007) Dictionary and introduction to global environmental governance. Earthscan, London. (1st edition 2007; other editions 2009, 2013)

Schmeier S (2010) Governing international watercourses-perspectives from different disciplines: a comprehensive literature review. Hertie School of Governance-Working Papers No 53. http://dx.doi.org/10.2139/ssrn.1658899

Toonen T (2011) The (changing) role of national government in multi-level (water) governance. In: van de Valk MR, Keenan P (eds) Principles of good governance at different water governance levels. Papers presented at a workshop held on 22 Mar 2011, Delft. https://www.unesco.nl/sites/default/files/dossier/water_governance.pdg_.pdf?download=1. Accessed 19 Dec 2017

Tropp H (2007) Water governance: trends and needs for new capacity development. Water Policy 9 (Suppl 2):19–30. https://doi.org/10.2166/wp.2007.137

UN WWDR UNESCO World Water Assessment Programme (2003) WWDR1: water for people, water for life. www.unesco.org/new/en/natural-sciences/environment/water/wwap/wwdr/wwdr1-2003/

UNESCO-GEF Initiative (2015) Groundwater governance: a global framework for action-A shared global vision for 2030

UN Water Thematic Paper (2008) Transboundary waters: sharing benefits, sharing responsibilities. https://www.issuelab.org/resource/un-water-thematic-paper-transboundary-waters-sharing-benefits-sharing-responsibilities.html. Accessed 19 Dec 2017

Varady RG, van Weert F, Megdal SB, Gerlak A, Abdalla Iskandar C, House-Peters L, with major editing by McGovern ED (2013) GEF-FAO groundwater governance project: a global framework for country action. Thematic Paper No. 5: Groundwater Policy and Governance. Global Environment Facility-FAO, Rome. 47 pp. http://www.groundwatergovernance.org/resources/thematic-papers/en/. Accessed 19 Dec 2017

Water Governance Facility (2011) http://watergovernance.org/water-governance/. Accessed 19 Dec 2017

Wijnen M, Augeard B, Hiller B, Ward C, Huntjens P (2012) Managing the invisible, understanding and improving groundwater. Water Partnership Program, Water Sector Board, World Bank, Washington, DC

Chapter 5
Transboundary Hydro-Governance

If all the world's peoples work together, a secure and sustainable water future can be ours.
Kofi Annan, Former UN Secretary-General

Abstract In this chapter, the international dimension is added to hydro-governance in order to define and analyse the so-called "Transboundary Hydro-governance". Waters shared by two or more different countries have to be managed equitably through cooperation between riparian countries: no matter if they form an international border or if they are crossing political boundaries, such waters are "Internationally shared" or "Internationally Transboundary" or simply "transboundary". Besides the UN General Assembly Resolution 64/292/28–67-2010 declaration recognising the universal human right to water, water belongs to humanity and cannot be appropriated exclusively by different groups or interests. Fair transboundary hydro-governance means that no-harm is produced from one country to other watercourse countries. In this chapter we analyse the various drivers of transboundary waters governance, to understand how such governance can answer the new challenges of economic changes world-wide, as the traditional concepts of governmental policy evolve with, among others, the growing understanding that public participation, in the various forms proposed in Chap. 4, is a major contributor to a "good" hydro-governance. We examine what international joint institutions can be established, defining their roles and modes of action, and their strengths and weaknesses, to prepare for Chap. 6 and the identification of the instruments that are needed to implement transboundary waters governance in practice.

Keywords Analytical framework · Drivers · Multidisciplinary · State · Non-state · Joint governing institutions

J. Ganoulis, J. Fried, *Transboundary Hydro-Governance*,
https://doi.org/10.1007/978-3-319-78625-4_5

5.1 The Cooperation Challenge

Most political and administrative structures and ethnic entities are human-designed, with the objectives of managing, protecting and controlling human societies by considering their culture and languages and their natural and economic resources. These political entities or countries do not generally have the natural boundaries of river basins and aquifers. While it is already rather difficult to manage surface water and groundwater within a given country, the culture and education differences, the varying economic priorities, the different historical heritages, sometimes even the different religious backgrounds that exist between nation-states contribute to complicate the governance and management of transboundary waters and the adaptation of political and human requirements to hydrological and geographical realities.

The most important issue and the real challenge for the governance of transboundary waters, both surface waters and groundwater, is resumed in the process for ensuring cooperation between riparian countries in order to decide the appropriate water allocation and effective management of shared water resources and in the same time avoid potential conflicts between neighbouring countries. For implementing a shared transboundary governance framework, cooperation between countries using transboundary waters should be established and improved with time, considering not only national legislations and policies but also international water laws and regulations. This will involve national and international institutions, economic and political instruments together with the participation of stakeholders and various interest groups. It is a multilevel framework, local, national and international, across political borders, which most often do not correspond to geographical and hydrological boundaries, already complicated for surface waters that can be seen and becoming quite complex for groundwater aquifers, due to their lack of visibility.

Ideally transboundary hydro-governance means defining long-term goals and agreeing on a long-term vision, setting up common monitoring networks or, at least, exchanging data and processing methods, identifying the spatial boundaries of the water bodies and agreeing over them, setting up joint institutional and operational structures and decision-making procedures for the integration of objectives and, if necessary, priority setting, ensuring transparency of water exploitation and protection caring for international constraints. It also means that governments of nation-states exchange their monopoly of national policy and power by adopting and using the principles of international legislation and negotiating a common decision making process.

As shown in Fig. 5.1, the best way for promoting transboundary hydro-governance through cooperation is to translate the "water sharing" issue to "sharing benefits" from different socio-economic water related activities, such as water supply and sanitation, agriculture, industry, livestock, tourism and recreation. An important result from cooperation is also the ecological preservation and the environmental protection from joint planning and protection measures.

From a practical point of view, planners and managers of transboundary waters on both sides of the boundary should consider joint planning based on joint studies and joint data collection or, at least, using data processing methods that yield

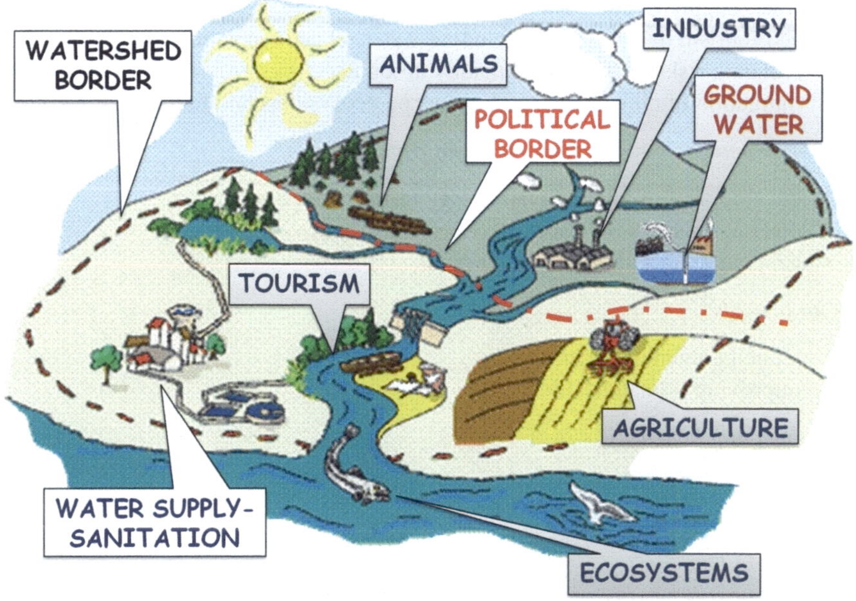

Fig. 5.1 Economic activities and potential benefits from transboundary water cooperation

compatible results. They should make the best of the opportunities created by remotely sensed data, including trust building. They should share information on the use of legal and financial instruments, e.g. abstraction permits or tradability among others and be prepared for emergency situations from extreme hydrological event, such as floods and droughts.

Although many countries, NGOs and UN organisations have made significant progress in several fields of transboundary water governance and management, such as data and information sharing, developing transboundary diagnostic analyses and suggesting effective water governance models, in many transboundary basins international cooperation at the basin level continues to be fragile.

Problems related to minority groups, perceptions of injustice and nationalistic tendencies continue to pose a threat to the stability and socio-economic development in many transboundary areas. It is therefore important to highlight the main qualitative and quantitative economic benefits that might be generated from an effective transboundary water cooperation.

Some of these benefits may be listed as follows:

Economic benefits within the transboundary river basin from:

- irrigated agriculture especially in the river delta regions: increase productivity while avoiding salt water intrusion from the sea,
- hydroelectric power generation: use of small dams in cascade while ensuring environmental protection,
- ecological flow rates and free flow of fisheries,

- water supply of cities and settlements,
- eco-tourism development, including water-related recreation,
- reducing risks of floods and mainly droughts,
- mitigation of climate change impacts on ecosystems and humans.

Economic benefits beyond the river basin from:

- development of international energy infrastructure networks,
- opening markets for goods, services and labour,
- generating cross-border investments.

Social benefits:

- reducing unemployment,
- improving human health.

Environmental benefits:

- preserving biodiversity,
- supporting migratory bird habitats.

Geopolitical benefits:

- reducing military defence spending,
- improving political stability.

In many regions around the world, water cooperation among countries is vital because a large amount of surface and groundwater resources are shared by two or more countries. As an example, in the Balkans, SE Europe, the Sub-Danubian transboundary river and lake basins and the internationally shared aquifers account for more than 80% of the water resources in this region (INWEB 2008). In this region, the UNESCO/ISARM program was initiated under the supervision of a scientific committee for monitoring and evaluation of the on-going cooperative activities. The UNESCO Chair/INWEB was assigned as the coordination institution. Milestone evaluation reports as well as presentations and results from workshops, seminars and stakeholders' meetings are publicly available in the INWEB Web site (www.inweb.gr).

5.2 Analytical Framework

5.2.1 Multidisciplinary Drivers and Actors

Governance and management of waters in transboundary areas are determined by different social drivers, i.e. entities that influence the way water is shared and used by riparian countries. These drivers are multidisciplinary, they are used by different actors and may refer to the laws, regulations and rights, supported by science and knowledge and stimulated by adequate economic and financial resources (Box 5.1). As will be presented in Chap. 6 next, social drivers can also be used as tools and instruments in order to improve existing hydro-governance.

Box 5.1 Social Drivers of Transboundary Hydro-governance
Social drivers are defined as social entities, institutions and agencies grounded in social norms and values, that determine directions and processes of hydro-governance change. Such divers are:

Data and information exchange/harmonization that serve to assess the current situation and perform a transboundary diagnostic analysis of existing water resources.
Sharing scientific and technical knowledge which is the base for understanding transboundary problems and suggest different technical, legal and socio-economic solutions.
Legislation and Policy that are composed by formal laws and informal norms—that shape the behaviour of people and affect their lives.
Socio-economic values, which may include respect to human rights and citizenship; commitment to social justice; tolerance; and respect for diversity.
Ecological issues, which deal with environmental security and biodiversity.

These drivers are implemented by institutions with a certain political commitment, assisted by scientists and technicians and stimulated by the involvement of different actors and stakeholders, which require good communication and awareness raising and the acquisition of information and knowledge through monitoring, education and training. Although these drivers may use existing multidisciplinary tools, which can be classified in different types, such as information, communication and awareness-raising, knowledge, science and technique, education and professional training, policy and legislation, socio-economy, they can be serve as instruments for improving the existing governance situation (see Chap. 6).

Hereafter, we will introduce and describe these drivers and examine how they should be adapted to the specificities of transboundary waters. As shown in Fig. 5.2, one very important driver for transboundary water governance is the *exchange and harmonization of monitoring data and information* available in riparian countries.

The assessment of transboundary waters relies on the availability and quality of the data, as in the case of any water body, but the transboundary aspects add other difficulties: on the one hand, the consistency of the data collected in different countries with different methods, and, on the other hand, the differences in interpretation of similar results due to differences in education and training (Fig. 5.2). It means that data should be harmonized—formats, classifications, terminologies, reference systems, levels and scales, specific software and hardware, among others, should be made compatible. The harmonization will also account for the differences in political, organizational, legal, cultural, and economical agendas and situations.

Data treatment will also depend on the desired presentation of the data, such as visualization, textual, web-based and real-time. Scientific data and knowledge, together with an appropriate legal and policy framework and the use of information

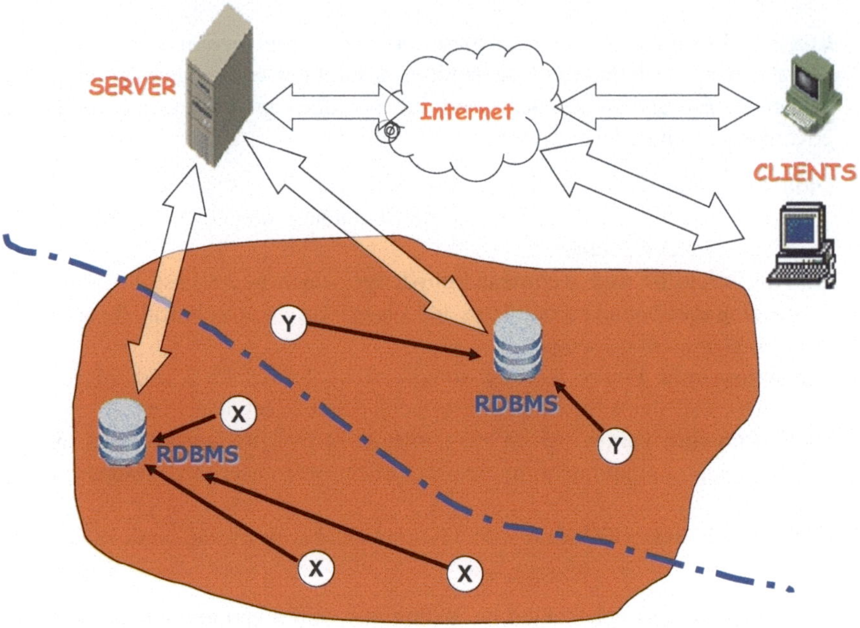

Fig. 5.2 Data monitoring and harmonising information in transboundary river basins/aquifers (RDBMS Relational Data Base Management System)

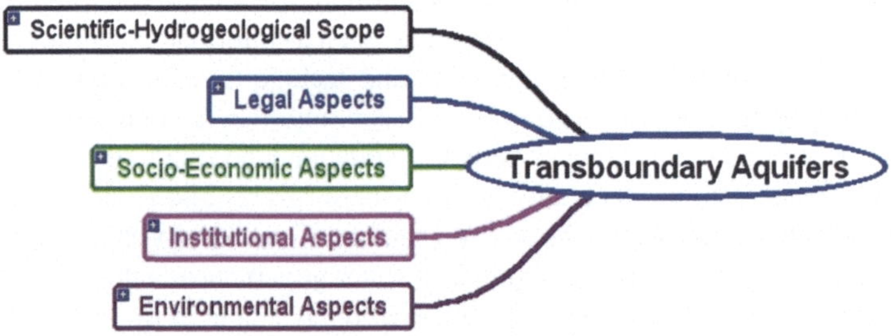

Fig. 5.3 Five focal areas of transboundary hydro-governance (ISARM 2001)

and communication tools, are the main drivers of the analytical framework of hydro-governance together with the interaction of different actors, such as transboundary institutions, NGOs, and voluntary compliance. The analytical framework should also integrate traditions and mental models, bureaucracy, conflict resolution mechanisms, and political economy defined as the study and use of how economic theory and methods influence and develop different social and economic systems.

ISARM (2001) has summarized five different social drivers/focal areas shown in Fig. 5.3 for transboundary groundwater governance, and, according to us, these are also valid for shared surface waters as well.

Exchanging scientific and technical knowledge is an essential driver for good water governance and management, starting from monitoring data and information processing, laboratory and field experiments in order to develop mathematical and/or physical models, change of scales, and, more generally ensuring the passage from theory to practice. Besides, good management requires good information and its diffusion. Decisions need to be based on proper knowledge and understanding of the general and local conditions; hence, it is important to ensure that sufficient data, information and knowledge are available and accessible to all. For example, the usual issue the decision-makers and their scientific staff must cope with is how to develop, use, control and protect the water resources of a given area optimally. It is evident that meaningful answers to this question cannot be given without having access to a basic suite of *data and information* on the local water systems (character, quantity, quality, recharge, development, uses etc.); the *setting* (socio-economic, ecological, political, etc.); and the *changes in governance processes* that will or may take place in the future. Knowledge based on sufficient and reliable data and information is thus essential to guide water exploitation, management and protection. Under transboundary conditions, what is needed are comparable data qualities and equivalent accessibilities on both sides (intercalibration of monitoring and data processing) because it is essential that this knowledge be available to both sides of the political boundaries. In many countries, such knowledge and the underlying data and information are insufficient or missing, even in relatively advanced countries, which may create significant discrepancies between countries sharing a water resource and having different states of knowledge. The state of knowledge, be it the types and qualities of data or the very state of science, as for instance what definitions are chosen, what modelling, either mathematical or physical, is used, how data are being processed, what interpretation is given, and how conclusions are transformed into decisions, may differ between neighbouring countries. This will add to the discrepancies across the border and will have to be taken care of for the sound management of a shared water resource. Adding to the issue, the differences in technical advancement may cause problem even if an agreement has been found for the management of shared waters.

Legislation and Policy are intimately connected, as laws and regulations are the major instruments of policy implementation and, usually, are strongly influenced by the type of policy that has been adopted. Policies define societal objectives and the institutions that will be used to reach these objectives, their mode of action and their role and responsibilities. Policies are completed by action plans, which set the various stages of policy implementation and their time-table and provide the framework for the drafting of the laws and regulations that will be needed to implement the policies. A water policy involves the vision of both its final objectives and the action plan to reach them, the sustainable commitment of the policy- and decision-makers in charge of water, and the understanding of the links between water and the sectoral economic policies where water plays a part. The implementation of a water policy relies on the sustainable commitment of the policy-makers which is the necessary condition for the long-term financial support. To be *aware of* and *knowledgeable about* water in the area concerned, its potential, the current and potential threats, and

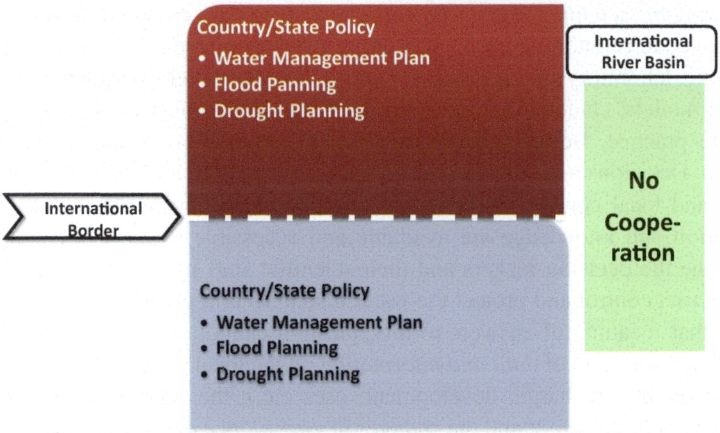

Fig. 5.4 Non-cooperative transboundary water governance

how to address these are essential conditions of good leadership. The ability to deliver on the action plan means the *capability* to put water management onto the agenda, to identify the main issues to be addressed, to mobilize essential stakeholders and to organize the overall water management process, with positive interaction amongst actors. It also means to be able to raise sustainable political support, which is the compulsory basis of financial support, and requires clearly presented policy messages based on reliable information. If national policies and water plans are implemented in transboundary basins without cooperation, then the non-cooperative model shown in Fig. 5.4 is far to be efficient.

A water policy and its action plan are implemented with the support and the codification of laws, regulations and rights. To quote Smith et al. (2016), whose paper deals with groundwater but can rightly be applied to water in general, "the law sets the rules for the exploitation and protection of water resources. A narrow, archaic view is that law establishes the means of enforcing rules through the courts. While this is necessary, law on water should create a stable framework of rules that enable governments and water users to plan water management over the long term. The role of law in water management is to put in place regulations in which the competing interests of all users of water can be fairly taken into account, including the interests of the environment and future generations". The laws and regulations are prepared, presented and adopted according to the institutional and legal organization of the concerned countries. The statute laws and regulations may be completed by existing rights and customary laws, i.e. un-written norms borne out of long-standing practice, which can be sometimes more binding on the members of the community who observe them. These customary laws are usually found in rural areas with long traditions of water management related to the long-time rules and habits of land and water ownership. Typical themes addressed by the legal module are water ownership and user rights, water protection from pollution, the rights and

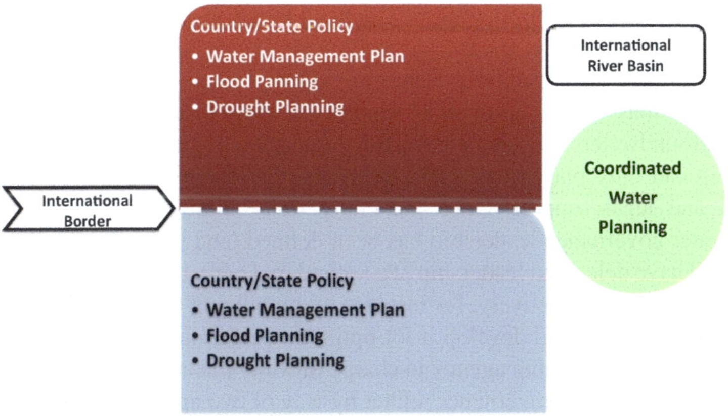

Fig. 5.5 Coordinated transboundary water governance

obligations of the different actors. Administration, performance monitoring, and eventual enforcement, of the laws are important facets of legislation for the implementation of policies and action- oriented management plans.

On both sides of a political boundary, laws and regulations are specific to the concerned country with the exception of the very few international legislation, essentially the 1992 UNECE Convention on the Protection and Use of Transboundary Watercourses and International Lakes (Water Convention), the 1997 Convention on the Non-Navigational Uses of International Watercourses and the 2008 Resolution of the UN General Assembly on the law of transboundary aquifers including in its annex the draft articles prepared by the UN ILC (International Law Commission), which "encourages the States concerned to make appropriate bilateral or regional arrangements for the proper management of their transboundary aquifers, taking into account the provisions of these draft articles".

Formal or informal transboundary cooperation between neighbouring countries may lead to a coordinated model of transboundary water governance, as shown in Fig. 5.5. Also international legislation has been at the origin of several agreements and conventions linking a few countries within the same basin, such as, among others, the 1994 Convention on Cooperation for the Protection and Sustainable Use of the River Danube or the 1999 Convention on the Protection of the Rhine. Because of the specificity of legislation, it could be the task of the joint management agency or team to work on the consistency of the laws that apply to the shared water bodies on both sides of the boundary and, perhaps, propose adaptations at national levels to be then adopted according to each country legal system. Particularly, where applicable and feasible, institutions, including private parties, should be legally obliged to make their data and information publicly available on both sides of the boundary, notwithstanding political and economic priorities, if the countries concerned really want a good management of their common waters. Generally, in a given country, a government agency is entrusted with leadership regarding water management, thereby receiving an appropriate legally defined mandate and resources to fulfil the corresponding tasks. The assigned government agency may be affiliated with a ministry focusing on a specific water-using sector (such as public water supply, or irrigation) or may be a 'user neutral' agency for water that can be completed by a specific groundwater management agency. An agency entrusted with the overall leadership can only play its role effectively if it has sufficient capacity as well as political and legal support. Once the entity to be entrusted with water and/or groundwater governance leadership has been defined (and possibly also the entities that should have delegated leadership at a secondary level), then due attention should be paid to support it effectively, for instance by mobilizing the constructive engagement of stakeholders, and develop it for optimum performance.

This concept of a leading agency in charge of water management will also be part of transboundary water governance, either by way of two agencies, one per country, establishing a joint team to manage the shared waters, or by way of a joint agency created by the water authorities of each riparian country. This has been the solution adopted for international river basins, such as the Rhine or the Danube. For groundwater, it has been the solution for a rather limited number of cases, such as the Geneva Aquifer Commission or the Joint Authority for the study and development of the Nubian Sandstone Aquifer, with varying responsibilities. From a practical point of view, the ability to conceive and commit to a clear vision and action plan and the capacity to deliver on that plan are very general requirements that will help identify the representatives or the politico-administrative entity of each country that will oversee the shared water resources. To these political qualities, diplomatic qualities should be added concerning both the managers and the people in charge of the negotiations, i.e. charisma and capacity to open to the others and ability to concentrate on their problems and understand them, both technical problems and political priorities, and curiosity to study and understand the specific political contexts of the negotiating parties. Finally, such agencies, which we introduce in a later chapter as River Basin Organizations (RBOs) and International River Commissions (IRCs), (Fig. 5.6) will fully rely on the team work capacities of their staff, stimulated by adequate education and professional training.

The costs of joint water governance and management should be evaluated and the financing mechanisms should be identified. Hence, socio-economy plays a major role in water governance and management, in several ways, through economic policies such as public-private participation or aligning incentives and subsidies systems for an efficient and sustainable use of water, through economic instruments, such as the polluter-pays principle, or which relate to the costs and benefits of a water policy, especially when it is transboundary and has to be shared by two or more riparian countries, through financial instruments such as the pricing of water or billing systems or the role of international donors, and, finally, through the involvement of stakeholders and the balance between their interests. For example, adequate water pricing acts as an incentive for the sustainable use of water resources and the price charged to water consumers, for instance for the supply of fresh water and the collection and treatment of waste water, should reflect the true costs. But, although

Fig. 5.6 Joint transboundary water governance by an International River Commission (IRC)

economic considerations are required for sound water governance and management, this can be quite delicate under transboundary conditions, because the state of development and the economies of the riparian countries can differ, involving differences in priorities that may not be always compatible. This is where the participation of stakeholders, including the general public, becomes most significant, through their social distribution and the nature of their involvement, the way they influence policy and management, or political economy, and, in general, the influence of Civil Society. In water governance and management, there are many stakeholders, often with conflicting interests, whether they belong to a given country or are located on both sides of a political and/or administrative divide.

Water is interlinked with many facets of every-day's life (water supply, energy, food, industry, land use, ecosystems, the environment, mining, use of the subsurface). In turn, each water system is exposed to many independent actions, by numerous individuals and entities, be it in the form of water supply and distribution, groundwater withdrawal, drainage, wastewater disposal, land use practices, irrigation, food production, energy production or use, transportation, polluting activities, nature conservation or environmental protection activities. More specifically for groundwater, the private sector is particularly important, because much groundwater use and pollution is related to private behaviour, ranging from farmers to beverage companies and other major consumers. All these individuals and all these entities are stakeholders, but their stakes are of a varied nature, and their demands and interests may be competing or even conflicting. Companies, for instance, may face a range of different obligations—social responsibility, reducing risk exposure, shareholder pressure etc. A particular challenge is to reach out to smaller stakeholders and to devise stakeholder processes that are inclusive, in order to ensure that the interests of poorer people are represented.

Good water governance should attempt to define and establish a sustainable 'best compromise' between the varying demands and interests, which requires the different categories of stakeholders to articulate their demands and preferences on either side of the boundary. It is for this reason that involving the stakeholders actively in different stages of the water management process will potentially lead to better policies and local water action plans. It will enhance their understanding of the compromises to be accepted and it will help align individual behaviour with the common goals adopted for local management and the measures selected for its implementation. Stakeholder involvement can come in many forms and intensities. Variants run from a rather passive involvement, where stakeholders are simply informed on what the government is going to do, to the other extreme where stakeholders are more or less self-regulating without significant interference of government agencies. Variants-in-between may include stakeholders being consulted, or involved as co-developers of plans or as actors with delegated implementation tasks. City mayors for instance can take a leading role and have done so in some fast-developing cities. Those acting as leaders in water governance should define which stakeholder participation model fits best the local situation, and how this model may evolve over time.

Stakeholders' involvement is not necessarily more difficult in transboundary conditions than inside a given country, despite language, education and cultural differences, as professional interests and priorities are often the same on both side of a boundary. But clearly two barriers have to be overcome: the language barrier and the cultural barrier. A good system of interpretation and translation should be set, with official translators and interpreters from both sides, and short intensive training courses on cultural aspects should raise the awareness of the various stakeholders on specific cultural aspects that may or will influence the governance and management of a shared water resource. This will complement the current awareness-raising programs and targeted communication events, which should exist in each country to support stakeholder involvement or stakeholder participation and ensure a consistent partnership approach over time. And we come back to a feature we have already introduced, a permanent, jointly mandated and managed agency, which should efficiently support such programs and events. Part of the transboundary water governance effort will be to define such an agency and its objectives, its composition in terms of staff both scientific and administrative, its location, its operation rules, and its relationships to the governments on both sides of the boundary.

The implementation of policy and legislation requires money, which will depend on political will and, therefore, will need the informed involvement of the decision-makers supported by their constituency, the public in general and the various stakeholders, which in turn implies that good information, communication and awareness-raising are provided. Usually two types of information are considered: the first type comprises a 'static' description of the systems considered, surface water systems and groundwater systems and the interlinked physical environment and human communities. Such information is usually collected during assessment studies and is presented and accessible in the form of reports, maps and databases. The second type captures dynamic, time-dependent variables by monitoring activities such as time-series of levels, quality parameters, volumes withdrawn, demography, and others for both surface water and groundwater. The rapid development of new techniques, including remote sensing, can make a difference in providing real-time and impartial information, and, of course, should be made available on both sides of the border. Such information, mainly scientific, is related to communication and awareness raising. Awareness of the nature and challenges of water is essential to guide decision makers, thus enabling informed decisions to be made, and to motivate all stakeholders to align their behavior with the objectives of good water management. Awareness will result from the dissemination of information and its conversion into knowledge, for which professionals with relevant expertise can be very useful. Sophisticated methods and techniques may be needed to analyze the related aspects, but the answers should be presented in a form that makes the messages understandable for those addressed (decision-makers, planners, local stakeholders, the general public). The analysis carried out and the data used should be documented in a degree of detail that enables other professionals to verify and—if necessary—expand the analysis, which requires solid education and training. Differences in education from one country to another may be an issue and when they share water resources, they will have to find solutions to these differences, such as, for instance, joint training sessions.

For achieving maximum impact and reducing costs, the data and the results of their interpretation and analysis should be shared as widely as possible with all stakeholders. This requires an open-minded attitude from the 'owners' of data and information, and the implementation of effective data-/information sharing activities. The latter may include negotiations amongst different entities to exchange data; the development of web pages and on-line databases; reports targeting specific groups of actors; dedicated publications in journals; brochures and other publications; articles and messages in newspapers, on radio and television, or disseminated via other mass media. Where applicable and feasible, institutions dealing with water, either public or private, should be legally obliged to make their data and information publicly available.

To summarize, four main tasks can be distinguished:

1. collecting data; storing, processing, interpreting and analyzing information;
2. sharing the acquired data and information;
3. translating and disseminating the results of interpretation and analysis in the form of tailor-made messages that give guidance to optimal development,
4. use, management and protection of water resources and the interconnected environment and ecosystems.

Under transboundary conditions, these tasks are more difficult to undertake, especially if collection and processing methods differ from one country to another. Intercalibration of data will be needed as well as an adequate preparation of the human resources, who will be in charge on both sides (WISER 2017). In a given country, a dedicated agency or department entrusted with all information and knowledge tasks concerning water, and often specifically groundwater, under one roof, and with full government support, is likely to contribute to these tasks being carried out systematically and coherently rather than in a fragmentary and haphazard fashion. We mentioned such an agency in the previous paragraphs, and even its possible name as River Basin Organization or Transboundary Basin Organization. Therefore, a similar agency could be established jointly between the countries sharing the water resources, whether surface water or groundwater, with the full support of the respective governments.

The various drivers of transboundary water governance are related and we have seen that they act in an integrated way. This integration is emphasized by a "horizontal" driver, horizontal because it concerns the various aspects of governance, from science and knowledge to legislation, to socio-economy and communication and awareness-raising: education and professional training. A major driver of water governance and management in general, education and professional training are still rather empirical when coping with transboundary situations. Fried (2016) has written that more systematic transboundary water governance education and training was needed because of the very significant role of transboundary water resources worldwide, stressing the need "to educate and train competent practitioners, such as water managers, hydrogeologists and related scientists, policymakers and diplomats". Fried emphasizes that the complexity of transboundary water governance is amplified, on the one hand, by foreign policy considerations often tending to use water resources and their management issues as a political instrument, and, on the other hand, by misunderstandings related to cultural and language differences, which leads to the introduction of another discipline, besides the classical science, law and socio-economy, "water diplomacy" or the methodology for negotiations and conflict resolution. From a practical point of view, it should be noticed that education and training need funding which takes us back to the driver "socio-economy".

Knowledge, information, communication and awareness depend on the nature of water: although few people in general know what is beyond the faucets of their bathrooms or kitchens, surface water is rather correctly understood by the public and by the decision-makers, which is not the case of groundwater. Therefore, **groundwater** may have to be dissociated from surface water in terms of awareness and awareness raising, which we explain hereafter.

Few people have a basic understanding of groundwater: its invisible presence inside geological formations below ground surface, its dynamics, its potential and the threats it is exposed to. Even fewer people have any knowledge of the local groundwater conditions in the area where they live. However, most people depend on groundwater in one way or another, and the majority of them should be aware about what is at stake, either because they have to take decisions on groundwater withdrawal, use, control and protection, or because they are users or polluters who need to be motivated to align their behaviour with the objectives of groundwater management. Awareness is thus a key component of good groundwater governance for all stakeholders, such as decision-makers, well owners, and water consumers. Examples based on the knowledge of local groundwater conditions can be used to develop tailor-made messages for awareness-raising among different categories of stakeholders. In particular, awareness of decision-makers should be raised to such a level that they experience a sense of urgency for groundwater management and are able to take well-informed decisions, while planners need to be enabled to develop realistic plans and to compare the pros and cons of alternative management strategies. Local stakeholders and the general public need to understand why certain management measures are being implemented, to what extent they are supported by legal frameworks, how individuals can contribute by changing their individual behaviour, what benefits can be achieved by concerted and consistent groundwater management action, and what problems may arise when refraining from any groundwater management intervention.

Transboundary conditions do not change or complicate awareness-raising actions. Whether the aquifers are in one country or shared by several countries, the awareness-raising actions are quite similar because the needs of each category of actors are usually defined by the type of activity of the actor and not by its citizenship: for instance, farmers on both sides of a political boundary face the same preoccupations of drought, flood, fertilizers or pesticides. The differences will be in the existing national and local legislations and how they can be applied in a consistent manner to the shared aquifers more than in the professional habits of the users, and this is where awareness-raising actions will have to be undertaken in complement to those mentioned above and, of course, not limited to groundwater.

Water resources governance in internationally shared river basins and transboundary groundwater aquifers can be effective if it is based on cooperation, peaceful resolution of potential conflicts and the development of a common vision between riparian countries. In its publication “Factors enabling transboundary aquifer cooperation-a global analysis” (Conti 2014), IGRAC has proposed to group the drivers of cooperative and conflictive transboundary water interactions into five broad categories (ecological, economic, geopolitical, legal/institutional and social).

Ecological drivers are a potential source of cooperation, especially IWRM which requires stakeholders from various sectors and levels of governance to manage surface water, groundwater and related ecological resources jointly in an integrated way across political boundaries, yielding benefits, for instance in terms of improved water quality or enhanced ecosystems services. And benefit-sharing by the riparian countries could stimulate them to cooperate more thoroughly instead of competing for water, especially when scarce. IGRAC stresses that exchange of information, joint monitoring and assessment, increased availability of scientific information will contribute to make ecological benefits more transparent and, hence, drive cooperation more efficiently. Besides, it has been noticed that water stress and water scarcity do not generate or increase conflicts and affect cooperation willingness (Wolf et al. 2003; Wolf and Fried, personal conversation 2017), confirming the potential role of ecological drivers in cooperation.

Socio-economics is another driver which stimulates cooperation through benefit-sharing beyond the water resources themselves, looking at the various sectoral economic policies and their relationships to water, such as energy through hydropower or geothermal energy production, agriculture, transport or tourism, and which, of course, influence neighboring countries. Social drivers act essentially by public participation and the role of civil society, which appear in international or national texts and laws, such as the Rio Declaration of 1992 for example, or even much before at national level as in the French Water Law of 1964. For transboundary situations, IGRAC gives the example of the Franco-Swiss aquifer of Geneva managed at local community level, which we present in Chap. 7.

Geopolitical issues affecting neighboring countries will clearly have consequences on their cooperation in the domain of water and the risks of conflicts. As underlined by IGRAC, it is not the type of existing political organization of the countries that affect their relationships concerning water but the possible significant political changes, such as the fragmenting of a country into smaller political units, and noticing that cooperation in non-water issues does not necessarily predict cooperation with respect to water (Wolf et al. 2003). Besides, as also underlined by IGRAC, power asymmetries between riparian countries or a unilateral disproportionate action by a riparian country within a shared basin or aquifer system could also create a conflicting situation in the domain of water. Geopolitical considerations can drive to the setting of a transboundary basin institution to preserve the long-term water interests of the riparian countries through cooperation instead of competition.

Several international rules and agreements have been adopted over the rather recent years as legal instruments for hydro-governance, also introducing international institutional mechanisms for governance, such as the UNECE Water Convention or the UN Watercourses Convention, among others. They are supposed to be efficient legal/institutional drivers of transboundary water interactions but it has been noticed that they did not fully succeed in that respect, due to weaknesses of the agreements or power differences between the concerned countries, and some researchers have asked for a redesign of these laws with a better anticipation of possible conflicts.

5.2.2 *Sovereignty Issues: State and Non-state Actors*

A key factor to be considered is the sovereignty of States, especially in the case of shared water resources, because the state is the referent level that is entrusted and empowered to make sovereign decisions over the management, use and conservation of these shared waters. In relation to water, state sovereignty is upheld/mitigated through the existence of a global institutional framework comprised of customary international water law (the norms dictating how states behave), global and regional conventions and bilateral or multilateral basin-level agreements and treaties and basin management organizations. This involves implementing and maintaining a structure of relations among nations guaranteed by procedures of cooperation and agreement and, as we will see in the design of an effective model of transboundary hydro-governance (Sect. 8.2), the creation of a transboundary basin organization is a way for the riparian countries to manage their shared waters while preserving their sovereignty. The UN estimates that approximately 3600 joint institutions exist between states to govern transboundary water resources (Earle and Neal 2017).

Yet a major problem in shared hydro-governance is the decrease or transfer of sovereignty that joint management of a resource will imply, but this will be achieved through negotiations, which can be long and tedious. It will be considerably facilitated when considering that transboundary hydro-governance also falls under the responsibility of non-state actors, as explained hereafter.

While transboundary hydro-governance is usually considered as being under the responsibility of the states with state actors being the key players in international waters negotiations and the implementation of agreements, it has been observed that water management and allocation decisions that are made at the international and national levels often rely on national and subnational organisations for their implementation and often rely on civil society and/or local communities for their acceptance and legitimisation. The inclusion of non-state actors in the decision-making analytical framework of transboundary water governance seems, therefore, imperative (Earle and Neal 2017). According to Suhardiman and Giordano (2012), these non-state actors are comprised of international financial institutions, international/transnational and domestic NGOs, civil society groups, and the private sector, among others. The non-state actors differ greatly one from the other, and this is well illustrated by the various existing NGOs, either international or domestic, with different objectives and modes of operation and addressing different levels of government. Besides, the border between State and non-State actors is not always clear and over-lapping happens, actors sometimes playing on both sides, which, of course, becomes even more complicated under transboundary conditions. Suhardiman and Giordano (2012) call such actors "hybrid" with an ability to change positions and organizational linkages to fit their changing interests and strategies. Quoting Suhardiman and Giordano, "The hybridisation processes reflect not only the increasingly complex challenges in transboundary water governance, but it also brings to light the ongoing multi-party governance process to cope with these challenges. Crucial in this process is the way state and non-state actors create a nested institutional structure, devising strategies, and coordinate management efforts at multiple governance level (from local to basin wide) as in the case of Chesapeake Bay, Great Lakes, and Baltic Sea programmes (Karkkainen 2002). The notion of hybrid actors reveals an additional layer of complexity in the hydro-hegemony concept (Zeitoun and Warner 2006) and contributes to our current understanding of power relations and interplays in two ways. First, it unravels the dialectic relationship between key actors' changing interests and the actual shaping of the rules of the game in transboundary decision-making processes. Second, it implies the temporary nature of transboundary decision-making outcomes and consequently bends the rigid notion of power asymmetry. Zawahri and Hensengerth (2012) illustrate this blurred boundary between state and non-state actors, highlighting the relationship of domestic NGOs leaders and government officials in China".

It should be stressed that the recognition that, first, states are not monolithic and should not be treated as a "black box" with a unified interest but as complex structures with organs, e.g. ministries or secretariats, having their own interests and agendas, and, second, that other actors than state representatives have a say in hydro-governance, will contribute to the taking into account of the various sectoral economic policies at state level (usually represented by the corresponding ministries), and the participation process of the regional and local communities, and, more generally, the public to hydro-governance, favoring more deliberative and transparent decision-making processes.

Concerning the negotiations that have to take place to build the agreement that should govern the management of the shared water resource, Suhardiman and Giordano (2012) introduce an interesting paradox of transboundary hydro-governance: a treaty, always rather rigid and without much room for changes and adaptation to evolving conditions, is not necessarily the best answer to the complexity and evolution of transboundary water governance to cope with the riparian countries development agendas. And they illustrate their comment with the example of the Mekong River Commission, whose "institutional analysis shows that the overall logic in Mekong transboundary governance is rooted in the missing linkages between national- and regional-level decision-making landscapes and processes. Here, the missing linkages should not be perceived as an impediment in transboundary decision-making processes, but rather have become part of the structural requirements that shape the overall logic of Mekong water governance". This flexibility keeps the dynamics of transboundary hydro-governance, therefore decreasing the risks of conflicts through the possible adaptation of the relationships between the riparian countries to new hydro-economic or hydro-climatic conditions.

This paradox of transboundary hydro-governance without a formal treaty is illustrated by the case of the transboundary aquifer of San Pedro between the US and Mexico and described and commented in the final report of the Binational Study of the Transboundary San Pedro Aquifer (Callegary et al. 2016), which essentially emphasizes the science and data harmonization drivers. Hereafter are some of its most relevant considerations:

> The United States and Mexico share waters in several basins that cross the international boundary. But there is no treaty between the two countries regarding groundwater management for the transboundary aquifers. Hydrogeological information is needed to understand the behavior of this complex binational hydrological system, but most scientific studies have been completed at a national or state level, typically without knowledge or consideration of the hydrogeological conditions on the other side of the border. Scientific knowledge and binational data on these transboundary aquifer systems could be beneficial to decision makers in both countries, serving as a consistent baseline to support the decision process.
>
> When an aquifer system is divided by an international border, coordination with scientists from both countries is desirable to enable a comprehensive understanding of the current system conditions to facilitate future management and administration of transboundary groundwater resources.
>
> To facilitate transboundary scientific cooperation and coordination, the binational Transboundary Aquifer Assessment Program (TAAP) was officially launched on August 19, 2009, with the signing by the Mexican and U.S. Principal Engineers of the IBWC (International Boundary and Water Commission) of the "*Joint Report of the Principal Engineers Regarding the Joint Cooperative Process United States-Mexico for the Transboundary Aquifer Assessment Program*". This document serves as the framework for U.S.-Mexico coordination and dialogue to implement transboundary aquifer studies. The document clarifies several details about the program such as background, objectives, roles and responsibilities, funding, relevance of the international water treaties, and the use of information collected or compiled as part of the program. Sharing data and information in

both countries creates the basis for cooperation and a common understanding in support of sustainable use and protection of water resources and achieving sustainable economic development. In addition, the intellectual and interpersonal interaction improves the relationships needed to coexist in harmony at a local and international level.

The technical team for this study of the Arizona-Sonora aquifers included personnel from the two federal governments (CONAGUA and USGS) and two state universities (University of Sonora-UNISON and University of Arizona-UA) working together under the auspices of the IBWC. The participants in this Arizona-Sonora effort also included technical expertise from the faculty and staff of the Geology Department at the UNISON and the UA Water Resources Research Center, enabling a significant advance in the binational understanding of the basin. The participation of the two universities incorporated an academic aspect that reflected a culture of transparency and openness common to academia around the world. This academic collaboration brought flexibility to the information exchange and the development of a framework of mutual understanding that benefits both federal governments.

Considering the basin characteristics described in this study, binational information was available to estimate the parameters of the regional climate and vegetation system, but there were no previously existing binational integrated maps with information for soils, geology, geophysics, piezometry, hydraulic parameters, or water quality. Differences on both sides of the border in hydrography, hypsometry, and terrain slopes were, at least in part, caused by differences in the resolution of digital elevation models. One of the important results of the study was to create binational maps and data sets for many of the data types needed for a regional hydrological analysis. The surface geology and geophysical information—electromagnetic, gravimetric, magnetic, and seismic—have been used to develop the conceptual model of binational hydrogeology. The available data for climate, hydraulic parameters, piezometry, and hydrogeology may be used to update a groundwater flow model. This process has also served to identify gaps in data and the importance of monitoring on a binational level. Notably, the information on soil types could not be integrated on a transboundary basis because each country uses a distinctly different classification system. This joint, uniform integration would require a complex, long-term project requiring bilaterally negotiated methods and the participation of a binational field team specializing in soil mapping. With this exception, the integration of transboundary hydrogeological data was possible although sometimes requiring joint interpretation and dialogue among scientists from both countries.

5.3 International Governance Institutions

5.3.1 Joint Institutions

For implementing effective transboundary water resources governance or shared hydro-governance, the creation of joint political institutions, often called International River Basin Organizations (IRBOs) or Transboundary Basin Organizations (TBO) is recommended. At national level, the establishment of RBOs is the main institutional tool stated by the EU-Water Framework Directive in order to apply the IWRM process. We will examine the structure, legal base, decision-making processes and other particular conditions needed to create effective international institutions. Historically, the example of the Tennessee Valley Authority (TVA) in the USA, although operating in a non-international context, is frequently cited as the prototype for enhancing not only successful water resources governance between different states, but also for developing sustainable socio-economic development in the region. In fact, water is directly connected to different economic activities, such as food, energy, industry, tourism and is the basis for environmental protection and ecosystem preservation and water governance is the way water can facilitate the development of all these economic sectors, through an efficient use of the joint institutions either national or transboundary. The usual types of Transboundary Basin Organizations, as proposed by Hooper and Lloyd (2011) within a report on IWRM (Integrated Water Resources Management) in Transboundary Basins, supported by the UNEP-DHI Centre for Water and Environment, are the following:

> An **advisory committee,** which generally refers to a formalized or quasi-formal organization in which individuals take responsibility for undertaking action planning and provide advice; governments may 'hand over' strategic planning to such organizations; they frequently have no or limited legal jurisdiction. An example of an Advisory Committee is the Intergovernmental Coordinating Committee (CIC) for La Plata Basin, in charge of promoting the harmonious development and physical integration of the River Plata Basin and its zone of direct and measurable influence, promoting, in the region of the Basin, the identification of areas of common interest, carrying out studies, plans, and works, and formulating such operating arrangements and legal instruments as are deemed necessary to achieve a series of objectives given in the agreement.
>
> An **association** (or initiative), which generally refers to an organization of like-minded individuals and groups with a common interest, similar to an Advisory Committee. In a river basin, they have varying roles: providing advice, stimulating basin awareness, education and ownership of basin natural resources management issues; and fulfilling educational functions and information exchange. An example of an Association is the Nile Basin Initiative, without any agreement, aiming at achieving sustainable socio-economic development through the equitable utilization of, and benefit from the common Nile Basin water resources.
>
> An **authority**, which generally refers to an organization which makes planning decisions at a central or regional government level. It may set and enact regulations, or have development consent authority. Authorities are founded on democratic principles and a framework of law to which all relevant individuals and institutions are subject in a basin setting. An example of an Authority is the Autorité du Bassin du Fleuve Niger, taking care of the integrated development of the basin notable energy, water resources, agriculture, animal rearing, fish breeding, sylviculture, transportation and communications, industry, and equitable sharing in the development, conservation and use of Niger Basin water resources.
>
> A **commission**, which generally refers to an organization which is delegated to consider natural resources management matters and/or act on those matters. A basin commission's powers vary, and include advisory/education roles, monitoring roles, undertaking works, fulfilling goals of a specific government's charter or an international agreement. Commissions normally are instituted by a formal statement of a command or injunction by government to manage land and water resources; commissions may also have regulatory powers". For example, a Commission is planned by the bilateral agreement between Moldova and Ukraine in the Nistru (Dniester) River basin, signed in 2016, hence posterior to the report. The new Treaty identifies principles and provides a framework for cooperation on water pollution prevention and control, water flow regulation, conservation of biodiversity and protection of the Black Sea environment. It also addresses the monitoring of data exchange, public participation and cooperation in emergency situations. A bilateral Dniester Commission is planned to be established, to facilitate sustainable use and protection of the Basin and encourage countries to develop and implement joint and coordinated Dniester River basin management plans. (Bilateral Agreement, Nistru River Basin 2016).

Another interesting example of a Commission with regulatory functions is the Interstate Commission for Water Coordination of the Aral Sea Basin, in charge of the regulation of water allocations, the definition of common water policy, the determination of rational use, annual consumption limits and water allocation, the organization of ecological programs for desiccation and determination of the Aral Sea requirements, the production of recommendations for common price policies and compensation for losses, the coordination of large scale projects, the creation of an information base, the coordination of joint research, the establishment of cooperative technical links.

According to the report, the usual functions of the Transboundary Basin Organizations are collecting and disseminating information and data, and the mediation of water use conflicts, with limited experience in recovering the costs of water supply and management (considered as being the role of national governments), and some environmental functions, such as monitoring and assessment.

5.3.2 Strengths and Weaknesses of International Joint Institutions

The SWOT Analysis (Strengths, Weaknesses Opportunities, Threats) presented by "*Integration of Groundwater Management into Transboundary Basin Organizations in Africa—a Training Manual*" (Vaessen and Brentführer (eds) 2015) is quite instructive and hereafter we give its main results concerning the strengths and weaknesses of Transboundary Basin Organizations (TBO's), which although established for Africa are according to us sufficiently representative of the general situation of a TBO:

(a) **Strengths:**

Governance:

- TBOs tend to operate within the framework of a multi-state agreement that provides for the possibility of a transboundary water management mandate.
- TBOs generally have a permanent secretariat that can initiate and manage programs and projects such as transboundary groundwater management and monitoring activities.
- TBOs can provide a platform for mobilizing basin-wide political support and for bringing groundwater higher on the political agenda.
- TBOs have a good platform for raising finance to carry out transboundary groundwater actions such as monitoring.

Riparian State Collaboration:

- TBOs can be a focal point for improved political and technical collaboration between riparian states.

Data Management and Sharing:

- TBOs can provide a suitable platform for hosting transboundary groundwater data and for the management and use of the data.

Capacity Development:

- TBOs can optimize groundwater management capacity development as a focal point for basin wide training programs in transboundary groundwater management.
- TBO's can optimize capacity utilization by providing a platform for the pooling of scarce technical expertise from the riparian states.

(b) **Weaknesses:**

Governance:

- Many TBOs are advisory bodies only and do not have a legal mandate to manage transboundary groundwater.
- Most TBOs have a strong traditional focus on surface water management and hardly consider groundwater.

Riparian State Collaboration:

- Many TBOs are not well integrated with the groundwater management authorities in the riparian states.
- There is often insufficient understanding of transboundary groundwater issues in TBOs.
- Disparities in groundwater challenges and context as well as groundwater development and management give rise to different focus between states.

Data Management and Sharing:

- Many TBOs have neither groundwater data, nor trained staff, nor a suitable computer platform for a groundwater database at present.
- Agreements on data sharing are often limited and ill defined. Data sharing protocols are often non-existent and riparian states may be reluctant to share their data under such circumstances.
- Most TBO agreements do not include any legal requirement for states to share their groundwater data, even in transboundary aquifer situations.

Capacity Building:

- Many TBOs do not have the skills, personnel, or the equipment to carry out transboundary groundwater management.
- There is often a lack of interest to develop groundwater technical capacity in TBOs, due to their focus on surface water resources.

References

Bilateral agreement between Moldova and Ukraine in the Nistru (Dniestr) River Basin (2016) https://www.slideshare.net/OECD_ENV/bilateral-agreement-between-moldova-and-ukraine-in-the-nistru-dniestr-river-basin

Binational Study of the Transboundary San Pedro Aquifer (2016) Final report. International Boundary and Water Commission, Universidad de Sonora, CONAGUA, University of Arizona, USGS

Conti K (2014) Factors enabling transboundary aquifer cooperation-a global analysis. IGRAC, Delft

Earle A, Neal MJ (2017) Inclusive transboundary water governance. In: Karar E (ed) Freshwater governance for the 21st century, global issues in water policy, vol 6. Water Research Commission, Pretoria

Fried J (2016) Introduction. In: Fried J, Ganoulis J (eds) Transboundary groundwater resources-sustainable management and conflict resolution. Lambert Academic Publishing, Saarbrücken, pp 1–7

Hooper BP, Lloyd GJ (2011) Report on IWRM in transboundary basins. UNEP-DHI Centre for Water and Environment, Horsholm, 40 p

INWEB (2008) Inventories of transboundary groundwater aquifers in the Balkans. UNESCO Chair and Network INWEB, Thessaloniki. http://www.inweb.gr

ISARM (2001) Framework document internationally shared (transboundary) aquifer resources management, their significance and sustainable management. https://isarm.org/sites/default/files/resources/files/framework.pdf

Karkkainen BC (2002) Post-sovereign environmental governance: the collaborative problem solving model. In: Biermann E, Brohm R, Dingwerth K (eds) Global environmental change and the nation state. Postdam Institute for Climate Impact research, Potsdam, pp 206–216

Smith M, Cros K, Paden M, Laban P (eds) (2016) Spring – managing groundwater sustainably. IUCN, Gland

Suhardiman D, Giordano M (2012) Process-focused analysis in transboundary water governance research. Int Environ Agreements 12:299–308

Vaessen V, Brentführer R (eds) (2015) Integration of groundwater management into transboundary basin organizations in Africa. Training manual by AGW-Net, BGR, IWMI, CapNet, ANBO, & IGRAC

WISER (2017) Water bodies in Europe, integrative systems to assess ecological status and recovery, Intercalibration. http://www.wiser.eu/background/intercalibration

Wolf A, Yoffe S, Giordano M (2003) International waters: identifying basins at risk. Water Policy 5 (1):29–60

Zawahri N, Hensengerth O (2012) Domestic environmental activists and the governance of the Ganges and Mekong Rivers in India and China. Int Environ Agreements: Politics, Law Econ 12 (3):269–298. ISSN 1567-9764

Zeitoun M, Warner J (2006) Hydro-hegemony – a framework for analysis of trans-boundary water conflicts. Water Policy 8(5):435

Chapter 6
Instruments and Tools for Shared Hydro-Governance

Into the well from which you drink do not throw stones.
Arabic Proverb

Abstract The practical implementation of hydro-governance for transboundary waters, or shared hydro-governance, needs a set of instruments and tools, belonging to, on the one hand, classical domains such as science, economy, society, legislation both national and international, policy, and, on the other hand, less classical such as diplomacy for conflict resolution like mediation and negotiation tools. We will more specifically explore the roles under transboundary conditions of joint monitoring, cost-benefit analysis, International Water Resources Management (IWRM), legislation, more specifically international law, and policy. We will conclude the chapter by a survey of conflict resolution and its tools.

Keywords Monitoring · Cost-benefit · IWRM · Legislation · Policy · Negotiation · Remediation

6.1 Scientific and Technical Tools

Scientific knowledge is a fundamental driver of shared hydro-governance as it enables to describe water bodies, e.g. rivers, lakes, aquifers among others and understand how water issues can be dealt with and management decisions taken if a constructive relationship between science and policy can be worked out, which is even more true for waters depending on more than one jurisdiction. Armitage et al. (2015) have identified five key conditions supportive of effective science-policy interactions: "(1) recognizing that science is a crucial but bounded aspect in water resource decision-making processes; (2) establishing initial conditions and shared commitment among actors; (3) understanding that social or group learning processes linked to science-policy interaction are enhanced through greater collaboration; (4) accepting that the collaborative production of knowledge about hydrological and associated socio-economic change and institutional responses is essential to

J. Ganoulis, J. Fried, *Transboundary Hydro-Governance*,
https://doi.org/10.1007/978-3-319-78625-4_6

build legitimate decision-making processes; and (5) engaging boundary organizations and supporting informal networks of scientists, policy makers and civil society".

Transboundary conditions raise hydro-governance issues, essentially related to data collection and processing, impact assessments, homogenization of results and their interpretation or intercalibration, exchanges of information, elaboration of joint conclusions, transfer back of these conclusions and recommendations to the riparian countries and their adoption within the institutional and legal systems of the riparian countries. These issues should come within the jurisdiction of a Transboundary Basin Organization (TBO), which implies that its role should be clearly defined and accepted. It could play a constructive role in the coordination of joint research activities, and joint education and training to build the capacities needed for undertaking joint scientific actions and correctly implementing the conclusions of data processing and interpretation. Besides, the TBO should evaluate the changing physical conditions of the basin through its scientific and technical actions, therefore providing the scientific basis to possible adaptations of the existing agreements between the riparian countries towards the improvement of transboundary hydro-governance.

6.1.1 Surface and Groundwater Monitoring

A most important role of science is the collection and processing of data, with the possible use of modelling, whether physical or mathematical, and, for transboundary conditions, the communication and exchange of information, data and processing results between the riparian states or entities, contributing to make ecological benefits more visible on both sides of the boundary, hence stimulating cooperation. Joint surface and groundwater monitoring schemes will play a major role, operating in an integrated manner at the various political and administrative levels, such as international, national, regional, and local on the one hand, and, on the other hand, in a multidisciplinary manner by collecting and interpreting hydro-dynamical, chemical, biological, physical data and integrating the results in a consistent model of the concerned waters, to be used for management decision-making, all these operations taking place under the guidance of transboundary water governance to fulfill the objectives of transboundary water policy.

An interesting example of joint monitoring as a scientific basis for transboundary water governance is provided by the EU Water Framework Directive (EU WFD) and its guidance document n°7 "Monitoring under the Water Framework Directive", proposing an overall methodological approach to monitoring for the implementation of the WFD.

For surface waters, three types of monitoring are described: (1) surveillance, (2) operational, and (3) investigative monitoring

For groundwater, (1) chemical status, (2) surveillance, and (3) operational monitoring are required completed by a quantitative assessment of the available groundwater resource. The example of the WFD is presented in Sect. 6.4.2 "Policy instruments: the EU Water Framework Directive".

An interesting concept is used by the WFD, called "inter-calibration", a word we have mentioned above. The WFD requires the intercalibration of assessment methods to set consistent and comparable standards for rivers, lakes and coastal waters across Europe. Good ecological status should be comparable between countries. In other words, monitoring schemes need to follow the same standards and should lead to comparable results. The essence of inter-calibration is to ensure that the definition of ecological status in member countries' assessment methods for biological quality corresponds to comparable levels of ecosystem alteration. Comparable environmental assessments are of crucial importance for industry to receive equal treatment on environmental sustainability criteria at the European internal market, instead of heterogeneous assessment levels and protection measures.

6.1.2 Indicators of Good Hydro-Governance

Among the scientific and technical tools that enable to evaluate and follow the efficiency of hydro-governance, the Indicators on Governance, Management and Environmental Performance, proposed by Knieper and Pahl-Wostl (2016) are interesting to mention. These authors have listed the following indicators:

1. Water Governance (polycentric water governance i.e. the presence of multiple centers of authority and distribution of power along with effective coordination structures):
   - Integration of domestic water legislation
   - Distribution of functions, responsibilities and authority
   - Vertical coordination between government authorities
   - Horizontal coordination between government authorities
   - Involvement of local governments
   - Degree of centralization
   - Technical capacity and economies of scale
2. Water management
   - Response to water pollution incidents
   - Comprehensiveness of water quality monitoring
   - Presence of sound hydro-meteorological monitoring
   - Understanding of groundwater resources
3. Environmental performance
   - Aquatic biodiversity
   - Surface and groundwater quality

6.2 Economic Instruments

Cooperation in managing transboundary water resources can be difficult to achieve for various reasons, such as differences between countries in technical and scientific advancements, specifications for monitoring and data collection, information sharing readiness and capacity, lack of harmonization in hydrological and socio-economic approaches, differences in socio-economic and cultural levels, lack of trust, conflicting objectives and priorities in relation to their history, sovereignty issues, possible territorial claims, lack of political willingness and, not least, because property rights are often unclear and contested.

An economic process has been proposed by Sadoff and Grey (2002) to facilitate cooperation, potentially offering greater scope for underpinning equitable agreements between riparian countries or entities, which they called ‘benefit-sharing’. The idea, well presented in the Concept Paper prepared for the 4th SADC RBO Workshop (2010) *on benefit sharing in the context of transboundary water resources management and development (Southern African Development Community)*, is that if the focus is switched from quantities of water to the various benefits derived from water use and allocation—in multiple spheres, including economic, social, political, and environmental—riparian countries or riparian entities of a given country will correctly view the problem as one of positive-sum outcomes associated with optimizing benefits rather than the zero-sum outcomes associated with dividing water when this water is finite or capped, because a reallocation of water implies that what is gained by one riparian country or entity is lost in equal amount by one or more other parties. The riparian countries or entities losing water volumes in such scenarios are commonly reluctant to proceed to an agreement, for obvious reasons. A Positive-Sum Outcome (PSO—sometimes termed a ‘win-win solution’) is characterized by simultaneous improvements over time of the various aspects concerning all the riparian countries or entities sharing a transboundary watercourse or aquifer system. When water volumes are not limited or capped, i.e. if these volumes can be increased, a “win-win” solution involving water volumes can be obtained. The SADC concept paper has listed a set of methods which enable water volumes to be increased, such as, for instance, the reduction of losses due to leakage or illegal connections that affect many developing countries, wastewater treatment and reuse, desalination of brackish or marine waters, inter-basins transfers, the better use of Green Water, Blue Water (and a combination of Green Water and Blue Water through supplementary irrigation) or Grey Water and an improved knowledge of virtual water.

While shared benefits from use and allocation are a logical corollary of the most fundamental principles of international water law, there are still many controversies as to the conditions under which benefit-sharing could serve best as an approach. Recently, the approach has been receiving wider attention in the literature and is increasingly applied in various basins to enhance negotiations, stimulated by the assessment and communication to the public of different individual and mutual benefits. However, relatively little attention has been paid to the costs associated with benefit sharing, particularly in the long run.

Defined as "the process where riparian countries or entities cooperate in optimizing and equitably dividing the goods, products and services connected directly or indirectly to the surface water or groundwater bodies, or arising from the use of their waters" (Ramoeli 2010), benefit-sharing is rather well described by Soliev et al. (2015) for surface water, i.e. rivers. We extensively quote his text hereafter, and we think that it is adapted to groundwater by simply replacing "river" by "river basin and aquifers", implying they are managed conjunctively.

Soliev et al ask three questions: (1) What benefits are there? (2) How can they be shared? (3) What are the costs of achieving shared benefits? And hereafter is their answer: "Sadoff and Grey (2002) determined four categories of benefits associated with cooperation as *environmental* (Type 1), with increasing benefits to the river; *economic* (Type 2), with increasing benefits from the river; *political* (Type 3), with reducing costs because of the river; and *catalytic* (Type 4), with increasing benefits beyond the river. The main critique on the typology is its practicality (Qaddumi 2008; Tafesse 2001; Nicol and Mtisi 2003; Turton 2008) as well as weakness in prioritization or identification of entry points. The latter is addressed by Phillips et al. (2008) whose methodology (Transboundary Waters Opportunity-TWO-Analysis) helps to see areas of priority when brainstormed by riparians. Overall, most scholars agree on the typology (Sadoff and Grey 2002) as it covers the whole spectrum and allows distinguishing directions for cooperation".

"Further, Sadoff and Grey (2005) define "benefit sharing" as "any action designed to change the allocation of costs and benefits associated with cooperation". The term "any action" can be interpreted either as a hindering factor or as an enabling factor, since it broadens the range of processes beyond the water sector, including the other economic sectors. Sadoff and Grey (2005) acknowledge the fundamental principles of international water law—equitable and reasonable use—first established in the 1966 Helsinki Rules (The Helsinki Rules on the Uses of the Waters of International Rivers) and then codified in the 1997 United Nations (UN) Convention on the Law of the Non-Navigable Uses of International Watercourses. However, they propose the benefit sharing approach as an alternative. Dombrowsky (2009) disproved it as an alternative approach showing the importance of underlying *property rights* if mutual benefits are to be achieved and suggested that the approach could be rather complementary in certain cases. This is captured by a more specific definition suggested by Phillips and Woodhouse cited in (SADC 2010) as the process where riparian countries or entities cooperate in optimizing and equitably dividing the goods, products and services connected directly or indirectly to the watercourse, or arising from the use of its waters."

"Later, Sadoff et al. (2008), when explaining "fair sharing of benefits", refer to Article 6 of the 1997 UN Convention, which enumerates seven non-weighted guiding principles. Theoretically, this seems to translate the already existing dilemma of equitable distribution in the traditional (water volume based) approach into the benefit sharing approach. From a practical perspective, Sadoff et al. (2008) suggest learning from the actual practices derived from existing international treaties related to the management of shared water resources as a starting point of negotiations referring to the database of transboundary agreements developed by Wolf (1999). However, the authors admit that "the benefits derived from water development have generally not been shared equitably" (Sadoff et al. 2008). The approach seems to be rather future oriented focusing on ex ante conceptualization of possible options to facilitate cooperation.

More broadly, the idea of benefit sharing seems to replicate the mutual gains approach of the negotiation research introduced earlier (Fisher and Ury 1981). However, both strongly relate to and are based on the utilitarian concepts of game theory and welfare economics, particularly the problems associated to a Pareto improvement. However, unlike the game theory concepts, literature on both benefit sharing and mutual gains go beyond computing possibilities and show enthusiasm calling for creativity in problem solving, thinking beyond quantities, issues at the table, sectors involved, and assumptions. While encouragement for cooperation is supported by all means here, the question arises whether the increased emphasis to cooperate and achieve "yes" in a negotiation might overshadow or even cause some possible crucial negative consequences. Especially in a complex environment of shared water resources, broadening the basket and bringing in other, often as complex, issues, thus merging two or more complex resource systems, might easily lead to increased transaction costs by creating even a greater number of potentially conflicting interactions in a longer period.

According to *UN Water Transboundary Waters (*2008), riparian countries should focus first on optimizing the generation of basin-wide benefits, and secondly on sharing those benefits in a manner that is agreed as fair. The use of water, rather than the allocation of water itself, provides by far the best scope for identifying mutually beneficial cooperative actions. The perception by all countries that a cooperative basin development and management plan which maximizes overall benefits is "fair" is essential to motivating and sustaining cooperation. It is therefore important that consensus over basic entitlements is reached and that attention is paid to the differential distribution of costs resulting from the use of the water resources of the entire water body in question. It should be recognized, however, that due to the limited amount of overall available water in some cases, such decisions sometimes involve very difficult trade-offs and choices.

Payments for benefits (or compensation for costs) can be made in the context of cooperative arrangements. Downstream countries can be compensated, for example, for the creation and operation of additional storage capacity by upstream countries. This basin solidarity also might entitle upstream countries to share some portion of the downstream benefits that are generated, and thus share the costs of these practices. It is important, however, to apply a special approach to those benefits and costs that are not easily quantifiable or commensurable. Payment for ecosystem services (PES)—such as for flood mitigation, regulating run-off and water supply—is a new and still contested approach. Nonetheless, if implemented well, PES has the potential to be an environmentally effective, economically efficient and socially equitable tool for IWRM that can internalize environmental costs, broaden sources of finance and create incentives for environmentally friendly investments and behavior.

6.3 Shared Integrated Water Resources Management (S-IWRM)

National water development and management are strongly linked to sustainable and responsible development and, despite added difficulties especially concerning how neighboring countries understand development and view their socio-economic and environmental priorities, this is also the case for transboundary waters. Thus, an integrated approach favoring long-term and contingency planning is needed, building resilience into vulnerable systems, with an emphasis on increased diversity and flexibility, and, for transboundary waters management, based on regional cooperation principles, focusing on international river basins and aquifer systems. Such an integrated approach is Integrated Water Resources Management or IWRM.

Fundamentally, IWRM recognizes that the many different uses of the resource are interdependent. For example, while the high demand of water for irrigation and drainage contaminated by agriculture means less fresh water for drinking or industrial use, city and industrial waste pollutes rivers and related ecosystems. Additionally, decisions about whether and how the water in a river is used may balance protecting the fisheries and related ecosystems and, on the other hand, ensuring water is used for irrigation, drinking, and others.

Introduced at the 1992 Dublin Conference on Water and Sanitation, held to prepare for the 1992 UN Conference on Environment and Development in Rio de Janeiro, IWRM is based on the following principles, which have been internationally recognized:

- Freshwater is a finite and vulnerable resource, essential to sustain life, development and the environment;
- Water development and management should be based on a participatory approach, involving users, planners and policy-makers at all levels;
- Women play a central part in the provision, management, and safeguarding of water;
- Water has an economic value in all its competing uses, and should be recognized as an economic good.

As defined by the Global Water Partnership (GWP), IWRM is "a process that promotes coordinated and efficient development and management of water, land and related resources to maximize the economic and social welfare without compromising the sustainability of vital ecosystems". This requires a coordinated approach by industry, agriculture, the water-supply infrastructure, and others. It calls for a holistic management of surface and groundwater, implemented with the entire river basin in mind, which can be complicated when the basin is shared by two or more countries.

Numerous challenges are involved, such as continuous changes in people's demands and values and structural transformations in society and environment, not to mention climatic anomalies and other exogenous shifts. IWRM recognizes that the many different uses of the resource are interdependent, because freshwater is finite and vulnerable.

Moreover, the basic concept of IWRM involves the inclusion of a participatory process for decision-making. In that respect, USAID has introduced another definition which complements the GWP one: “A participatory planning and implementation process, based on sound science that brings stakeholders together to determine how to meet society’s long-term needs for water and coastal resources while maintaining essential ecological services and economic benefits, IWRM helps to protect the world’s environment, foster economic growth and sustainable agricultural development, promote democratic participation in governance, and improve human health”.

In transboundary areas, the IWRM framework should be ensured from both sides of the political boundary. If there is no cooperation or only limited coordination between riparian countries and stakeholders living in transboundary basins, conflicts on sharing and managing transboundary water resources can be expected. This is the case when hydrological extremes, as floods and droughts may affect the transboundary river basin or aquifer and one country may cause harm to another from flood events or experience painful water scarcity in case of long term low precipitation in the area.

Application of a Shared IWRM (S-IWRM) approach in transboundary hydro-governance is possible if riparian countries reach the political willingness to develop and operate a common planning of the transboundary water resources through cooperation.

A model of S-IWRM is shown in Fig. 6.1, where the central role at every particular step is interacting and consulting with all stakeholders living on both sides of the political border, independently of their formal nationality.

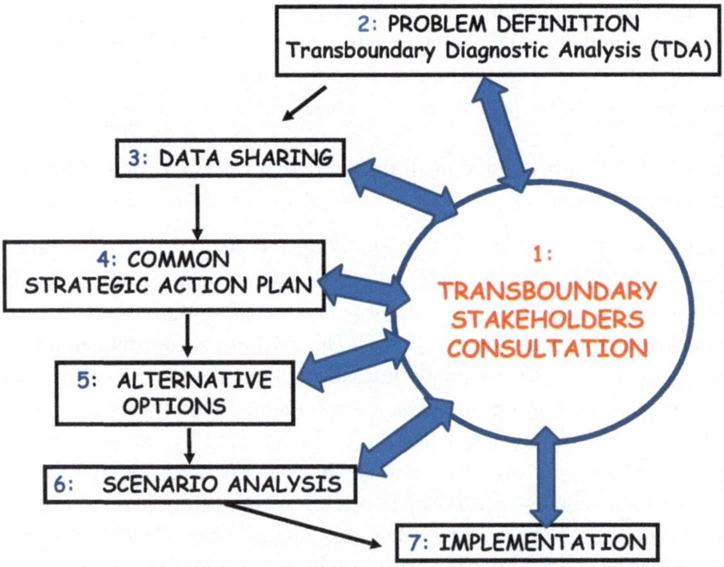

Fig. 6.1 A model for Shared IWRM (S-IWRM) in transboundary areas

The model consists of seven parts or steps, the first one being the continuous interaction and consultation. The other steps are as follows

1. Transboundary Diagnostic Analysis (TDA);
2. Data Collection, Information Sharing and Common Monitoring Systems (CMS);
3. Developing a Common Strategic Plan (CSP);
4. Produce Alternative Options;
5. Scenario Analysis and Shared Benefits Allocation (SBA);
6. Implementation: Transfer to Stakeholders, Applications at local scale.

It is important to underline that the old consultation model shown in Fig. 6.2a provides interaction after completing the water management plans. The new interactive and collaborative model shown in Fig. 6.2b starts the planning activities by incorporating stakeholders' participation at the early stages as well as by the end of the process during the implementation stage.

USAID underlines the following features of IWRM:

- IWRM is a coordinated process focusing on both economic and social issues and protecting ecosystems and ecosystem services, and:
- uses scientific (and indigenous) data, knowledge and tools to provide a sound base for judgment,
- emphasizes proper governance involving democratic participation through stakeholder engagement,
- recognizes that water is a finite resource and that different demands for water often compete, e.g. economic, social and ecosystem,
- finally, that water is cross-sectoral and thus that water management must be integrated across sectors.

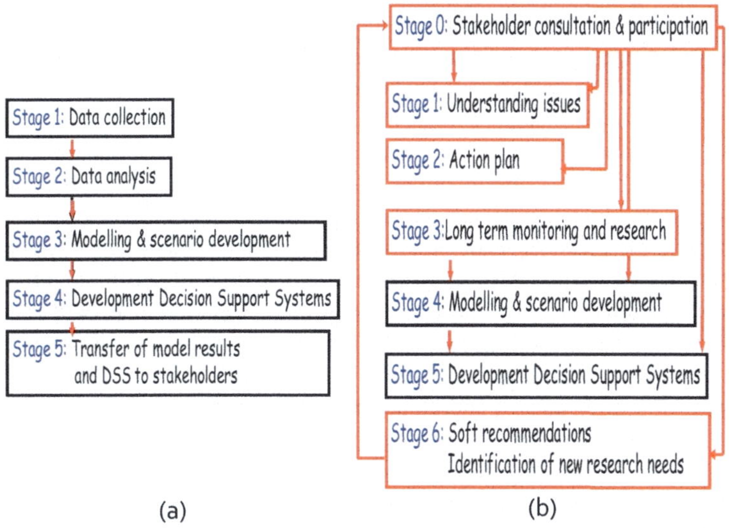

Fig. 6.2 The old (**a**) and the new model (**b**) prior stakeholders' consultation and feedback

USAID stresses that IWRM covers all parts of the water cycle, surface water and groundwater, as well as natural and anthropogenic interactions with and within it e.g. rainfall, catchment, runoff, recharge, quality, quantity, timing, use and demand, supply (*USAID-IUCN Ecosystem approaches and integrated water resources management-interrelated approaches-Fact Sheet N° 1*).

These definitions and features of IWRM certainly correspond to what water planners and managers need to guide their actions but they must not remain theoretical. Therefore, GWP has proposed instruments to implement these definitions and principles in practice. These instruments are listed and detailed in the following publication of GWP to which we refer our readers in the list of references, the IWRM Tool Box.

As an instrument of hydro-governance, how can IWRM be adapted to transboundary hydro-governance? As we can see with the GWP IWRM Tool Box, the implementation of IWRM calls for rather complex decision-making processes, which become even more complex for transboundary waters. Hence, at present, many existing transboundary cooperation arrangements are highly sectoral and the majority addresses specific waterworks, water uses and measures to control and regulate water flows, and others pollution or the environment. In order to fully follow the IWRM principles, there is a need to revise these approaches and, for instance, consider that the sectoral cooperation agreements should be actively used as the building blocks of an integrated approach, with the right mechanisms as well as changes in legislation. Globally agreed targets and indicators for IWRM plans are very important, since such plans provide an opportunity to assess the current water situation in all its thematic and subsectoral dimensions, as well as quality and quantity aspects. (*UN-Water Status Report on IWRM and Water Efficiency Plans for CSD-16* 2008). (*UN-Water thematic paper Transboundary Waters: sharing benefits, sharing responsibilities* 2008).

As introduced in Chap. 5 and presented in Chap. 8, a key transboundary hydro-governance component is a joint managing institution, the transboundary basin organization or TBO, which will be the driver of an IWRM plan. But there are various types of transboundary basins, hence this will imply that no standard IWRM plan can be applied in all circumstances. Consequently, the implementation of IWRM by transboundary basin organizations is a process with many variations according to the context relevant to the basin in question. That context includes differences in the physical features, critical water management issues (river pollution, water sharing, etc.), levels of economic development, institutional arrangements and natural resources management arrangements in each of the member countries (the 'riparian' countries), and of course the reason for the existence of the institution. In general, a transboundary basin institution has clearly identified roles and responsibilities as it coordinates action across boundaries and implements transboundary river basin management plans in response to changing conditions. It operates effectively within established institutional arrangements, uses transparent reporting mechanisms and is driven by stakeholder engagement. Besides it is not static and will evolve in time to respond to new situations, to answer new economic imperatives and to absorb new knowledge gained from science.

Once a transboundary basin organization exists, it will implement IWRM with the assistance of the riparian countries and the involvement of stakeholders of these countries. Different user groups (farmers, industrialists, entrepreneurs, communities, government institutions and NGOs, etc.) can and should be part of the process of defining strategies for water resources development and management. This provides additional benefits because trained and informed users are also involved in issues related to water conservation and watershed protection, water courses, aquifers, and others. Of course, it means that these users are trained to work together or, at least, made aware of the needs of the other groups and other countries. In this sense, the implementation of programs, policies and action plans to consider developing a water culture suited to the characteristics and needs of a specific community in each country and in riparian countries is a priority goal to ensure renewability provided by the natural water cycle, often conditioned on human actions. The pillars that support the way toward this goal are, without doubt, awareness, training and participation in the development of action-plans and decision-making processes, which should be directed to and supported by a clear knowledge and techniques based on natural and social processes involved in the dynamics of water systems, institutional strengthening and legal framework, directly relating to water governance and the search for efficiency in the use of water resources, moving towards a sustainable use of these.

The implementation of a transboundary IWRM plan should be followed to understand whether it is successful and contribute to good management of shared waters. Part of this follow-up is to evaluate the costs and benefits of the plan and how they are shared by the riparian countries and entities, which will depend on the economic instruments and principles used, such as, for instance, public-private partnership, polluter-pays principle or user-pays principle.

6.4 Legal and Policy Instruments

6.4.1 International (Global, Regional) and Domestic Legal Instruments

Two or more countries may consent at regional level to voluntarily produce a written document specifying certain cooperative commitments and obligations concerning the use of transboundary waters: in this case, we have *an international agreement* between riparian countries on transboundary water utilisation and management. Different names can be used for defining such a document, as international:

• treaty	• convention	• agreement	• understanding

These agreements should comply and be in accordance from the one hand with the *Domestic or National legislation on water resources* and from the other hand with the *Global UN Water International Law*. In fact, UN and UN Regional Commissions can suggest International Laws establishing transboundary

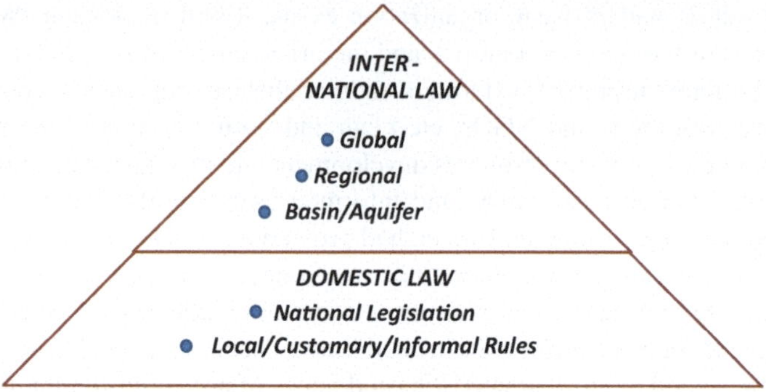

Fig. 6.3 International and Domestic Law at different scales

cooperation for water resources management for agreement and ratification by their Member States.

As shown in Fig. 6.3 conflicts and cooperation between countries can be regulated by two types of law: (1) the Regional or Global (UN) International and (2) the Domestic Law.

Legal instruments deal with regulatory provisions and legal duties of riparian countries in terms of access, utilisation, protection, preservation and management of transboundary waters. The codification of such legal rules is very useful to the international community, even though this process may be somewhat general and unable to cover the specific cases totally. The main challenge is whether different national administrations will agree to implement international rules at the national level and at the same time coordinate their activities with riparian countries through bilateral or regional collaborative agreements. This challenge may be faced by raising public and stakeholders' awareness in participatory processes involving national institutions, academic partners and international organisations, as emphasized in (Fried and Ganoulis (eds.) 2016).

Legal rules govern any society and are the bones which structure the World of Law. To determine the *legal conditions* that produce *legal effects* through *a legal rule*, we need legal definitions of real-life concepts (such as "pollution", "aquifer", "significant harm", etc.). It is important to understand that although these legal definitions should describe exactly the phenomena of real life in most cases this is not possible with great accuracy. Therefore, it is important to establish good communication and cooperation between lawyers and non-lawyers (such as hydrogeologists), so that legal definitions can match non-legal concepts and phenomena in the best possible way (Fig. 6.4).

Domestic or State law systems have a vertical structure with the central and federal State authorities at the top. In the World of Law, States are the original subjects since they possess sovereignty and supreme power. Thus, they exert this power on the persons and the things located in their territory.

The international law system, where all the Sovereign States are on the same level and standing on an equal footing, has a horizontal structure. Indeed, although the way the real world seems to be ruled contradicts this assertion, in the World of

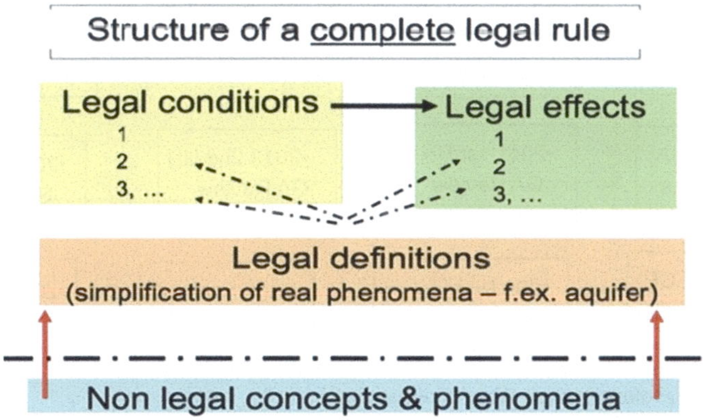

Fig 6.4 Legal and real-life (non-legal) concepts (Fried and Ganoulis (eds) 2016)

International Law, all the States are equal. Most international river basins do not have any legal cooperation mechanisms, yet legal instruments do exist and start to be implemented in several basins. But even then, this implementation is not always a success.

For surface and groundwater issues the following International Conventions/Directives/Laws exist:

At Global Level:

- The 1997 UN International Watercourses Convention (UN IWC)
- The 2008 Draft Articles on the Law of Transboundary Aquifers (UN General Assembly, GA)
- From 2016, the 1992 UNECE Helsinki Convention

At Regional Level:
Europe:

- The 1992 UNECE Water Convention and Draft Provisions on Transboundary Groundwaters
- The EU 2000/60 Water Framework Directive (EU/WFD) and
- the 2007 daughter directive on groundwater.

Figure 6.5 shows the main dates of establishing the Water International Laws/Conventions on internationally shared surface waters and aquifers.

The two significant UN legal instruments that provide important guidance on how states should approach the beneficial exploitation of their transboundary water resources are:

Fig. 6.5 Time evolution of the Water International Law

1. the 1992 UNECE Convention on the Protection and Use of Transboundary Watercourses and International Lakes (UNECE TWC) and
2. the 1997 UN Convention on the Law of the Non-Navigational Uses of International Watercourses (UNWC).

The 1992 UNECE Convention on the Protection and Use of Transboundary Watercourses and International Lakes (Water Convention) or *Helsinki Convention* was adopted in Helsinki in 1992 and entered into force in 1996. It aims to protect and ensure the quantity, quality and sustainable use of transboundary water resources by facilitating cooperation. It provides an intergovernmental platform for the day-to-day development and advancement of transboundary cooperation. Initially negotiated as a regional instrument (most countries sharing water resources and belonging to the United Nations Economic Commission for Europe or UNECE are Parties to the Convention), it turned into a universally available framework for transboundary water cooperation, following the entry into force of amendments in February 2013, opening it to all UN Member States. As of 1st March 2016, countries outside the ECE region can accede the Convention.

The Water Convention fosters the implementation of integrated water resources management, like the River Basin approach, which contributes to the achievement of the Millennium Development Goals and other international commitments on water, environment and sustainable development. It requires Parties to prevent, control and reduce transboundary impact not to cause significant harm, use transboundary waters in a reasonable and equitable way and ensure their sustainable management. Parties bordering the same transboundary waters should cooperate by entering specific agreements and establishing joint bodies, as mechanisms for the settlement of disputes. As a framework agreement, the Convention does not replace bilateral and multilateral agreements for specific basins or aquifers; instead, it fosters their establishment and implementation, as well as further development.

The 1997 Convention on the Law of the Non-navigational uses of International Watercourses complements the 1992 Convention as it details factors for equitable and reasonable water resources utilization while the 1992 Convention prescribes the contents of transboundary water agreements and the tasks of joint institutions for transboundary water cooperation. Its two key principles are "equitable and reasonable use" of water and "the obligation not to cause significant harm" to neighbors.

The 1997 Convention considers groundwater only if it is connected to surface water, as defined in its Article 2 "Use of Terms": *"watercourse" means a system of surface waters and groundwaters constituting by virtue of their physical relationship a unitary whole and normally flowing into a common terminus.*

The 1992 Convention states that "transboundary waters are any surface or ground waters which mark, cross or are located on boundaries between two or more States" (Tanzi 2014)

Through the accession to these two conventions and the acceptance of their binding rules, states will be called to fully cooperate on the utilization and the management of the aquifers and to solve peacefully their conflicts based on the principles and provisions stated. Then, another step to enhance trust and collaboration could be taken through the generalization of River Basin Commissions or joint bodies between riparian States which are not included within the conventions, but are strongly encouraged. Yet, at present, most states sharing the identified transboundary aquifers worldwide do not have adequate bilateral or multilateral agreements on shared waters, which may constitute a source of tension and generate risks for the sustainability of the resources (UN Watercourses Convention Online User's Guide). Besides, as stated by Qaddumi (2008) "Unfortunately, appeals to international water law have offered little direction on volumetric water allocations, partly because they themselves are contradictory and partly because propounded notions such as 'equitable use' and 'avoidance of significant harm' remain vague and, therefore, difficult to apply directly. History shows that riparians negotiate water arrangements according to their own methods, rather than appealing to principles of international law that are not universally accepted (e.g. Israel and Jordan on the Jordan river basin, Sudan and Egypt on the Nile river basin)".

The Case of Transboundary Aquifers: The UN Draft Articles on the Law of Transboundary Aquifers

In December 2008, the General Assembly of the United Nations adopted the draft articles on the law of transboundary aquifers. It addresses access to the water resource, reassuring the sovereignty of aquifer states over the portion of aquifer or aquifer system located within their territories. Yet their access to the resource is not unregulated but it is conditioned by the general rules of international law applicable to shared natural resources. The general obligation to cooperate should be promoted, use of the resource should be equitable and reasonable, with the obligation not to cause significant harm. A proper methodology for the regular exchange of data and information should be established by the aquifer states, and agreements could set up permanent transnational bodies or mechanisms to support the exchange of data and information. Monitoring is acknowledged as indispensable for the proper management of a transboundary aquifer. Ideally, states should reach joint monitoring based on an agreed conceptual model. States are encouraged to adopt management plans for the transboundary aquifers and aquifer systems, as well as to consult with the other aquifer state and establish joint management mechanisms (del Castillo 2016; Stephan 2016).

"The 63rd session of the UN General Assembly adopted Resolution **A/RES/63/124** on the Law of Transboundary Aquifers by consensus on 11 December 2011. The resolution encourages States 'to make appropriate bilateral or regional arrangements for the proper management of their transboundary aquifers, taking into account the provisions of these draft articles', which are annexed to the resolution. These provisions include cooperation among States to prevent, reduce and control pollution of shared aquifers. In view of the importance of these 'invisible resources', States are invited to consider these draft articles as a basis for the elaboration of a convention. The Law of Transboundary Aquifers is a concrete step forward towards the peaceful sharing of groundwater resources. Until then there was no instrument of international law that could provide a complete set of recommendations and guidelines for the sustainable and peaceful management of transboundary aquifers" (UNDESA UN Water, Water for Life 2014).

As we commented above, the implementation of existing international laws and agreements is not always successful, for reasons which usually are specific to each case. Yet Mianabadi et al. (2015) have managed to identify several broad categories of reasons, among which, besides hydrological or climatic differences, they list the absence of a comprehensive water law, the ambiguity of international water laws and non-integrated water management (the failure of IWRM).

6.4.2 Policy Instruments: The EU Water Framework Directive

Most Europeans live in transboundary river basins, therefore the European Union (EU) has chosen the River Basin approach as the best way to manage its waters and the best model for a single system of water management, a river basin being the natural geographical and hydrological unit, instead of the administrative or political boundaries. This approach constitutes the very basis of the major EU water legislation, the Water Framework Directive or WFD, which has established binding requirements for the management of transboundary watercourses for all EU members, increasingly complied with by non-EU-members in the region as well and is largely based on the setting of River Basin Organizations and River Basin Management Plans. First, let us examine the general requirements of the WFD and then their adaptation to transboundary cases, as presented in the following references (Schmeier 2010), (European Commission WISE "Introduction to the new EU water framework directive" 2016) and (Quevauviller 2016).

Member States initiatives concerning the Maas, Schelde and Rhine river basins have served as positive examples of the River Basin approach, with their cooperation and joint objective-setting across Member State borders, or in the case of the Rhine even beyond the EU territory. While several Member States have already adopted a river basin approach, it is not presently the case everywhere. For each river basin district—some of which do cross national frontiers—a "River Basin Management Plan" must be established and updated every 6 years. The presently existing (December 2016) International River Basin Plans are the following: Danube, Rhine, Elbe, Ems, Finnish-Norwegian International River Basin District, Meuse, Scheldt/l'Escaut, Odra, Sava Commission (ISRBC)

The purpose of the WFD and its daughter on groundwater is to establish a framework for the protection of inland surface waters, transitional waters, coastal waters and groundwater. More precisely the aim is to:

- Prevent further deterioration and enhance the status of aquatic ecosystems.
- Promote the sustainable water use based on a long-term protection of water resources.
- Enhance protection and improvement of the aquatic environment.
- Ensure the progressive reduction of pollution of groundwater and prevent its further pollution.
- Contribute to mitigating the effects of floods and droughts.

As shown in Fig. 6.6, evolution on time and the key elements of the WFD include various considerations, such as:

- Technical: monitoring, establishing the River Basin Management Plans (RBMPs).
- Institutional: adopt the river basin authorities as decentralized unit for water management.
- Environmental: protect water quality and ecosystems.
- Water economics: water tarification, cost recovery, environmental costs.
- Public participation: information, dialogue and public involvement.

The plan is a detailed account of how the objectives set for the river basin (ecological status, quantitative status, chemical status and protected area objectives) are to be reached within the timescale required. The plan will include all the results concerning the river basin's characteristics, a review of the impact of human activity on the status of waters in the basin, estimation of the effect of existing legislation and the remaining "gap" to meeting these objectives; and a set of measures designed to fill the gap. One additional component is that an economic analysis of water use within the river basin must be carried out. This is to enable a rational discussion on the cost-effectiveness of the various possible measures. It is essential that all interested parties be fully involved in this discussion, and indeed in the preparation of the whole river basin management plan, which involves a significant public participation, for two main reasons:

The first reason is that the decisions on the most appropriate measures to achieve the objectives in the river basin management plan, based on an economic analysis, will involve balancing the interests of various groups. Hence, the process should be open to the scrutiny of those who will be affected, which increases the adaptability of the plan to the societal changes.

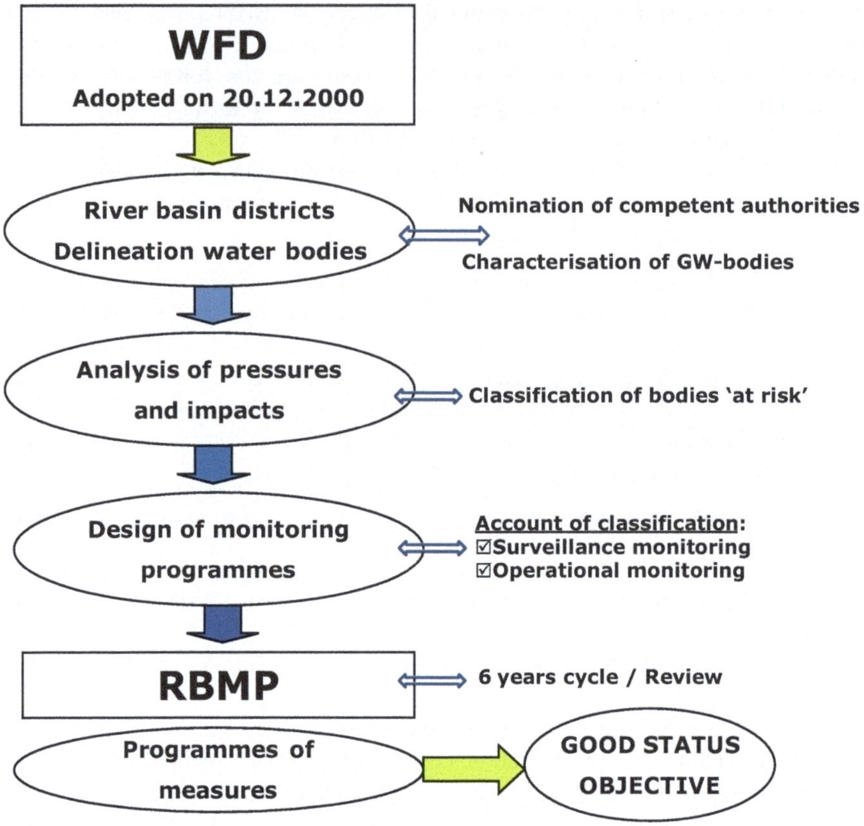

Fig. 6.6 Main elements and time evolution of the EU WFD

The second reason concerns enforceability. The greater the transparency in the establishment of objectives, the imposition of measures, and the reporting of standards, the greater the care Member States will take to implement the legislation in good faith, and the greater the power of the citizens to influence the direction of environmental protection, whether through consultation or, if disagreement persists, through the complaints procedures and the courts. The Water Framework Directive will require information and consultation when river basin management plans are established: the river basin management plan must be issued in draft, and the background documentation on which the decisions are based must be made accessible. Furthermore, a biannual conference to provide for a regular exchange of views and experiences in implementation will be organized. Too often in the past implementation has been left unexamined until it is too late—until Member States are already woefully behind schedule and out of compliance. The Framework Directive, by establishing very early on a network for the exchange of information and experience between water professionals throughout the Community will ensure this does not happen.

An important driver of governance introduced by the WFD is pricing and getting the prices right, which aims at answering the need to conserve adequate supplies of a resource for which demand is continuously increasing. Adequate water pricing acts as an incentive for the sustainable use of water resources and thus helps to achieve the environmental objectives under the Directive. Member States will be required to ensure that the price charged to water consumers—such as for the abstraction and distribution of fresh water and the collection and treatment of waste water—reflects the true costs. Whereas this principle has a long tradition in some countries, this is currently not the case in others. However, derogations will be possible, e.g. in less-favored areas or to provide basic services at an affordable price. In the case of transboundary waters this implies an agreement between the riparian countries, whether member States or not, involving financial aspects.

For transboundary groundwater bodies, specific features have been underlined (Quevauviller 2016). First, it should be noticed that the EU legislative framework defines "groundwater bodies" as reporting units, not to be confused with "aquifers" which may contain one or several groundwater bodies following the characterization and delineation process. Like river basins, aquifers may be shared by different countries and transboundary aquifer management has hence to consider several issues such as impacts of groundwater upstream abstraction on groundwater downstream flows and depletion of the groundwater table, as well as transport of dissolved contaminants with possibilities for transboundary pollution. In the context of the WFD, transboundary cooperation covers various issues, as follows:

Member States are required to establish appropriate coordination for transboundary river basin districts (including those extending beyond the territory of the EU), with the aim of achieving the WFD objectives throughout the entire river basin, integrating surface water and groundwater. The application of the rules remains with the Member States for the part of the river basin falling in their territory.

This coordination includes the characterization of pressures and impacts and delineation of groundwater bodies (not aquifers!). In principle, transboundary aquifers should lead to the delineation of international groundwater bodies requiring a coordinated management. It should, however, be noted that Member States may choose to split a transboundary aquifer into one or several groundwater bodies, making it a group of "national" groundwater bodies managed by a single Member State despite its international character.

Monitoring is directly linked to the characterization and delineation of groundwater bodies (at risk or not). In the case of transboundary groundwater bodies, one or more Member States should coordinate monitoring for the identified pressures.

The establishment of threshold values in principle should be coordinated for international groundwater bodies, including those extending beyond the EU territory (in other words, when threshold values are established at groundwater body level, they should cover the entire international groundwater body where such body has been delineated).

River basin management plans should be undertaken in principle for international river basins (one single plan per basin). Member States must publish groundwater status assessment at the level of river basin districts or parts of international river basins falling within their territory.

The key water governance concept underlying the WFD is "integration" as illustrated in the Guidance Document n° 7 "Monitoring under the Water Framework Directive":

- "Integration of environmental objectives, combining quality, ecological and quantity objectives for protecting highly valuable aquatic ecosystems and ensuring a general 'good' status of other waters;
- Integration of all water resources, combining fresh surface water and groundwater bodies, wetlands, coastal water resources at the river basin scale;
- Integration of all water uses, functions and values into a common policy framework, i.e. investigating water for the environment, water for health and human consumption, water for economic sectors, transport, leisure, water as a social good;
- Integration of disciplines, analyses and expertise, combining hydrology, hydraulics, ecology, chemistry, soil sciences, technology, engineering and economics to assess current pressures and impacts on water resources and identify measures for achieving the environmental objectives of the Directive in the most cost-effective manner;
- Integration of water legislation into a common and coherent framework. The requirements of some old water legislation (e.g. the Freshwater Fish Directive) have been reformulated in the Water Framework Directive to meet modern ecological thinking. After a transitional period, these old Directives will be repealed. Other pieces of legislation (e.g. the Nitrates Directive and the Urban Wastewater Treatment Directive) must be co- ordinated in river basin management plans where they form the basis of the programmes of measures;
- Integration of all significant management and ecological aspects relevant to sustainable river basin planning including those which are beyond the scope of the Water Framework Directive such as flood protection and prevention;
- Integration of a wide range of measures, including pricing and economic and financial instruments, in a common management approach for achieving the environmental objectives of the Directive. Programmes of measures are defined in River Basin Management Plans developed for each river basin district;
- Integration of stakeholders and civil society in decision making, by promoting transparency and making information accessible to the public, and by offering a unique opportunity for involving stakeholders in the development of river basin management plans;
- Integration of different decision-making levels that influence water resources and water status (these could be at a local, regional or national level), for an effective management of all waters;

- Integration of water management by different Member States, for river basins shared by several countries (*underlined by the authors*), existing and/or future Member States of the European Union".

To conclude this example, we think that the WFD presents the main features of the model of transboundary water governance we propose in Chap. 8, namely River Basin Organization and Planning, integrated surface water and groundwater management, significant public participation, intercalibration of data monitoring and processing, among others. Therefore, the WFD could be an inspiration for transboundary water governance worldwide.

6.5 Diplomatic Instruments

6.5.1 Alternative Dispute Resolution

In Chap. 3 different types of water conflicts have been analysed in terms of intensity, spatial extent and risk origin. As a terminology, we may agree that in increasing order of severity degree, the terms *dispute*, *conflict*, *armed conflict* and *war* will be used below (from UNITAR 2017). A further analysis of different causes of water related conflicts is summarised in Fig. 6.7.

Data Conflicts
Lack of data and unreliable information, contradictory information, or misinformation are frequently a cause of conflict. In transboundary river basins, different interpretation of the same data and different national standards for monitoring can contribute to disputes. There is a tendency to define most water conflicts or potential conflicts as data problems. The underlying assumption is often "if they only had better information they would understand and agree." However, better information could result in understanding better the conflict and it may not contribute to conflict resolution.

Relationship Conflicts
These are conflicts related to poor communication, misperceptions, personality differences, and stereotypes. This kind of conflict produces strong emotions and often must be addressed before people are able to resolve other forms of conflict. Sometimes this kind of conflict is resolved by increased communication or by getting to know each other better. The intervention of third-party is often needed to facilitate better communication.

Structural Conflicts
Structural conflict means that the general arrangement of the situation is set up in such a way that no agreement is reached. The "structure" that causes the conflict may be the way that roles and relationships have been defined, unreasonable time constraints, unequal power or authority, unequal control of resources, or

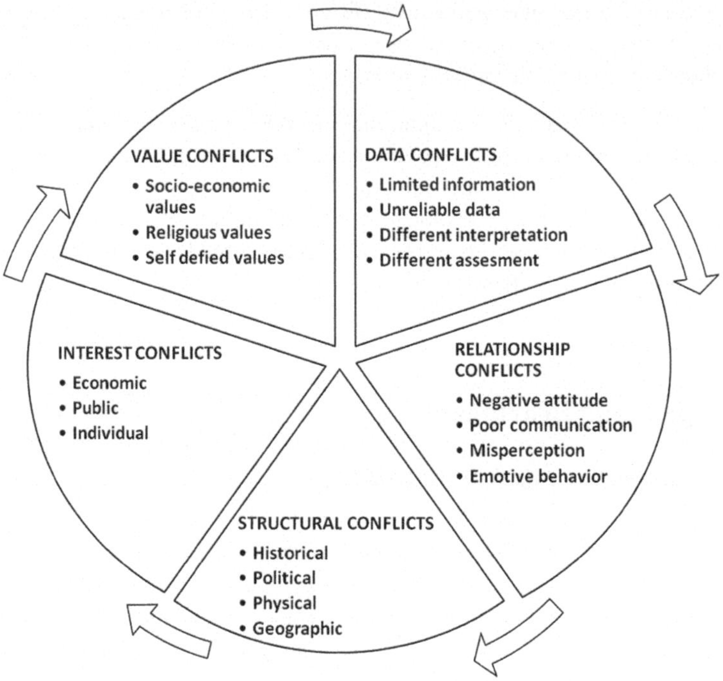

Fig. 6.7 Different causes of conflict (adapted from Delli and Wolf 2009)

geographical or physical constraints. Historical, political, physical and geographic reasons can also negatively influence a-priori the situation.

Interest Conflicts

Interest-based conflicts may be distinguished as

- substantive (physical resources, time, money),
- procedural (the way the negotiations take place), or
- psychological (trust, fairness, desire for participation, respect).

For a water dispute to be resolved, all parties must have a significant number of their interests addressed by the proposed resolution in each of these three areas. Often it is necessary to address data conflict or relationship conflict before addressing interest conflict. But if there are conflicts over interests, the dispute will not be addressed to people's satisfaction, until their interests have been addressed.

Values Conflict

Values conflicts occur when people disagree about what is good or bad, right or wrong, just or unjust. While people can live with quite different values systems, values disputes occur when people attempt to force one set of values on others or lay claims to exclusive values systems that do not allow for divergent beliefs. Resolution of values disputes sometimes occur, at least over time, as people educate each other about the basis for their beliefs. Beliefs about environmental values, for example, have changed considerably over the past two decades, at least in part because of this education process. Values conflicts can also be resolved when people build upon their many shared values, rather than concentrate on their differences. Or values conflicts may be resolved when the situation is structured so it is not necessary to resolve the differences.

Water disputes or conflicts are subject to transformation in terms of their intensity, which means that initial confrontations can be, like any other form of conflict, by nature very unique in the geographical, political and socio-economic context in which they take place.

There does seem to be some general patterns in approaches to water conflict and the way in which they are transformed, which have emerged over time. "Typical" disputes between, for example, developers and environmentalists, rural and urban users, or upstream and downstream riparians, suggest zero-sum confrontations where one party's loss is another's gain and where confrontation seems inevitable. Yet such intractable conflicts are regularly and commonly resolved, as creative thinking and human ingenuity allow solutions which draw on a more intricate understanding of both water and conflict to come to the fore.

Here is one path to the transformation of water disputes from zero-sum, inflexible disputes, to positive-sum, creative solutions, which centres on a migration of thought generally through five stages. Note that all stages exist simultaneously, and need not be approached in sequence, and no stage needs be achieved necessarily for "success". In today's world, many disputes never move beyond the first or second stage, yet are tremendously resilient, while a few have achieved the fourth stage and are fraught with tension. Nevertheless, like any skill, it is useful to understand the structure of an "ideal" path, in order to perfect the tools required for any individual situation. The generalised path described here, is structured around an understanding of each of the five stages through any of four perspectives, as described in Fig. 6.8

In stage 1, in its initial, adversarial setting, regional geopolitics often overwhelms the capacity for efficient water resources management. Metaphorically, the political

Fig. 6.8 Transformation of water conflict to active collaboration (adapted from Wolf 2010, UNESCO PC-CP, IHP)

boundaries on a map at this stage are more prevalent than any other boundaries, either of interest, sector, or hydrology. Dialogue is often focused on the past, based on the rights to which a country feels it is entitled, and a period of expressing pent-up grievances can be necessary. As a consequence of these initial tensions, the collaborative learning emphasis is on trust-building, notably on active and transformative listening, and on the process of conflict transformation. By focusing primarily on the rights of countries, inefficiencies and inequities are inevitable during this stage of negotiations.

As the adversarial stage of negotiations plays out, occasionally some cracks can be seen in the strict, rights-based, country-based positions of each side (although in actual water negotiations, this process can last decades). Eventually, and sometimes painfully, a shift can start to take place where the parties begin to listen a bit more, and where the interests underlying the positions start to become a bit apparent. In this stage 2, a reflexive stage, negotiations can shift from rights (what a country feels it deserves), to needs (what is actually required to fulfil its goals). Conceptually, it is as if we have taken the national boundaries off the map and can, as if for the first time, start to assess the needs of the watershed as a whole. This shift, from speaking to listening, from rights to needs, and from a basin with boundaries to one without, is a huge and crucial conceptual shift on the part of the participants, and can be both profoundly difficult to accomplish, and absolutely vital to achieve for any movement at all towards sustainable basin management. To help accomplish this shift, the collaborative learning emphasis is on skills-building, and we approach the (boundary-less) basin by sector rather than by nation. Once participants have moved in the first two stages from mostly speaking to mostly listening, and from thinking about rights to needs, the problem-solving capabilities which are inherent to most groups can begin to foster creative, cooperative solutions.

In stage 3, an integrative stage, the needs expressed earlier begin to coalesce together to form group interests—the "why" underlying the desire for the resource. Conceptually, we start to add benefits to the still boundary-less map, and in fact to think about how to enhance benefits throughout the region, primarily by adding resources other than water, and geographic units other than the basin. The collaborative learning emphasis is now on the consensus-building of the group, and we begin to move in "benefit-shed" rather than being restricted by the basin boundaries. Finally, while tremendous progress has been made over the first three stages, both in terms of group dynamics, and in developing cooperative benefits, Stage 4, the last action stage helps with tools to guide the sustainable implementation of the plans which have been developed, and to make sure that the benefits are distributed equitably amongst the parties. The scale at this stage is now regional where, conceptually, we need to put the political boundaries back on the map, reintroducing the political interest in seeing that the "baskets" which have been developed are to the benefit of all. The collaborative learning emphasis is on capacity-building, primarily of institutions. Active collaboration is the final fifth stage, comprising implementation, capacity increase and dissemination.

It is critical not to think of these stages as a linear process, where the further the better. Most basins ebb and flow back and forth over time, finding the level that meets a particular set of hydro-political needs for a given place and time—there is no "right" set of answers. One might think of these all existing in parallel "universes" simultaneously, each with its own set of approaches or tools, any of which may be useful at any given time, or conceptually as a helix or set of spheres rather than strictly linear. We break them apart here only for the purposes of explanation.

6.5.2 Identifying and Sharing Benefits

Although conflict is sometimes inevitable, it does not have to end in polarised disputes. Below are some of the positive externalities which can be brought about by conflict.

Conflict identifies problems that need to be solved—The first and most obvious value of conflict is that it identifies a source of dissatisfaction, irritation or anger for one or both parties. It also allows for a clarification of one's purpose: "What is important to us?"

Conflict is a way of creating evolutionary change, thereby reducing the need for violent upheaval—Change is an integral part of everyone's life: ideas change, and what we want from our relationships changes. The demands placed upon us by society, our jobs, families, friends, and spouse change around us, and the relationships must somehow accommodate change. When the necessary adjustments can be made in small steps, then a relationship can sustain considerable modification without threatening its stability. But if one or both parties deny or resist conflict, then the relationship becomes static, or like the dry limbs of a dying tree that are unable to bend in the wind. Without the ability to respond, recognise the conflict, and find solutions, the only alternative is either ending the relationship or confrontational demands to alter the relationship or institution, which can end in pain, even violence. Recognising conflict and responding to it in a positive way can prevent stagnation, allowing one or both parties involved to adjust the balance of power, and revitalise the basic values upon which the relationship is built.

Conflict helps us define who we are—The successful transition from adolescence to adulthood is a process of discovering the boundaries between parents and children. Adolescent rebellion is really a process of determining where "I" stop and "you" begin. Most adolescents solve this problem by testing their parents. Their parents' reactions, even their negative reactions, actually help adolescents establish their own sense of individuality, answering the question "How are we different (or the same)?" Similar issues exist in most close relationships. We seek out closeness because on a deep emotional level we experience completeness, a sense of being more fully ourselves when communication is open and intimate. But this very sense of connectedness can be a threat to our sense of individual identity. We may feel unbearably dependent on the people we care about and come to resent that dependency. When this occurs, we may find our separate identity by rebelling against the other person, which helps us get a sense of our distinct individuality. This process of "finding our outer edges", our boundaries, is particularly important in extremely intimate relationships such as those between parent and child or husband and wife. One or both parties may initiate conflict because they feel threatened by the closeness, and thus fear losing their individual identity. Much the same thing occurs between groups. Research has shown that groups need to "differentiate", to make their differences clear, before they are willing to focus on their common interests. Establishing ego boundaries, or group boundaries, appears to be essential to the health of both individuals and groups. However, the process should not stop with just establishing a boundary. A "mature" relationship acknowledges both differences and shared interests.

Conflict is a way of discharging some of the animosity or resentment that is generated by the limiting aspects of a relationship, or of sharing resources—Even the best and healthiest relationships place some restrictions on the individuals involved. When we are in a relationship, we cannot ignore the other person's (or group's) needs and ways of doing things. We put limits on our behaviour in order not to hurt other people, and hope they do the same. Although accommodation to another person's necessities can become excessive and unhealthy, some amount of accommodation is inherent in every healthy relationship. And even though we do not always want to admit it, this creates frustration and resentment. The frustration or resentment may become even stronger if we have to share resources or limit our behaviour because of individuals or groups we see as different from us. When others are of different ethnic background, religion, or beliefs, it is far easier to express the resentment, without the constraints of a caring relationship. Conflict provides a safety valve for expressing some of the inherent tension that results from these restrictions. When we have no way to express these feelings, we become increasingly aware of our frustration and resentment, and less aware of how much we care for the other person. By discharging the inherent tension, we become aware once again of shared interests and the desire for continuing relationships.

The bottom line here is that getting the best out of a conflictual situation depends on the way it is managed, and how positively it is managed. Therefore, one of the key jobs of any professional at any level of the conflict management command chain is to manage conflict so that it does not become dysfunctional, so that it does not evolve into a full-fledged dispute.

6.5.3 Reaching Agreements

Bi- and multi-lateral agreements on transboundary river, lake and aquifer water resources management are important tools for enhancing effective cooperation, involving political commitment and implementing joint water management plans. Depending on the legal status of the riparian countries, developing multilateral regional agreements has different degrees of difficulty. Three different cases can be distinguished:

1. Countries belonging to a *unified legal system* for water management, like the European Union (EU). Agreements are facilitated by the application of common water directives (e.g. the 2000/60 EU Water Framework Directive and the 2007/60 EU-Flood Directive).
2. Federal countries like the USA, Australia and Canada, may have more difficulties in developing inter-state water agreements as legal responsibility for water is given by constitution to the States, the Federal Government not interfering as coordinator.
3. Centralised countries may perform more easily international negotiations and setup legal issues for concluding agreements.

According to the previous experience of INWEB (International Network of Water/Environment Centres for the Balkans, 6th World Water Forum 2011), there are at least three main obstacles for developing and implementing multilateral agreements on shared water management:

1. lack of political willingness,
2. differences in socio-economic and cultural levels between riparian countries and
3. lack of trust and mutual interests, conflicting objectives and different priorities between countries, in relation to their history, sovereignty and possible territorial claims.

Although there are some good examples of bi- and multi-lateral agreements on transboundary river basins and aquifers, such as the Danube River Convention and the Guarani Aquifer Agreement, in many cases existing bi- or multi-lateral agreements on transboundary surface waters and shared groundwater aquifer systems are outdated, incomplete or obsolete. Many of them are not implemented in real terms or focus on very specific and particular issues, such as flood protection, navigation or hydroelectricity production, without referring to Integrated Water Resources Management (IWRM). Some are not in accordance with the principles of international law on transboundary water resources or do not rely on adequate national or common institutions and/or are not based on common monitoring and evaluation mechanisms.

There is therefore a need to increase the number of such agreements and to provide tools and methodologies in order to increase their quality and facilitate their implementation. In order to achieve this general objective INWEB has suggested the following recommendations in the 6th WWF:

(a) critically review existing bi- and multi-lateral agreements,
(b) screen examples of case studies which have been proven as successes in the field,
(c) examine the reasons for failures or incomplete agreements,
(d) develop indicators for good practice and effective implementation of mutual agreements and
(e) enhance the dialogue and consultation between members and stakeholders of the group in order to identify solutions, tools and methodologies for developing new agreements and/or improve the quality of the existing ones.

FAO has estimated that 3600 international water agreements or treaties have been concluded from AD 805 to 1984 for the purposes of governing uses of transboundary water resources and that number continues to grow. In the International Freshwater Treaties Database, University of Oregon, nearly 450 international, freshwater-related agreements, covering the years 1820–2007 have been identified covering 286 major Transboundary River Basins, compared to 5 international agreements only (!) for 445 Transboundary Aquifer/Groundwater Basins (Figs. 6.9, 6.10 and Table 6.1).

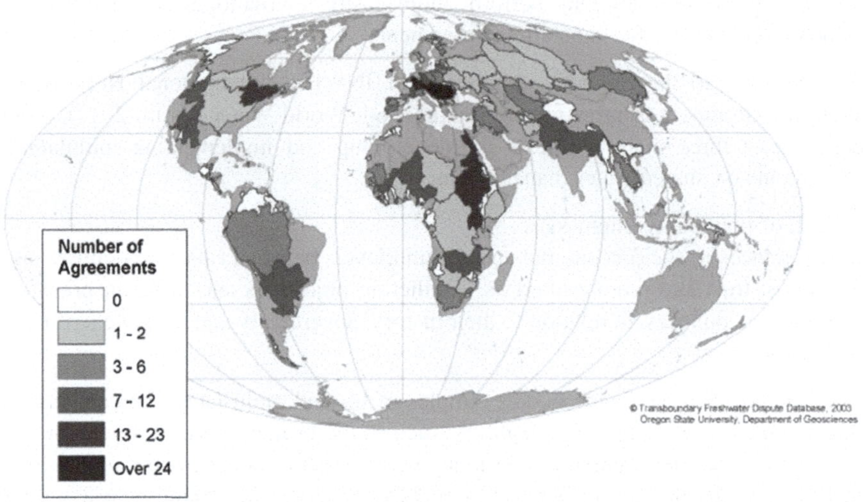

Fig. 6.9 International agreements per basin (Oregon State University 2018)

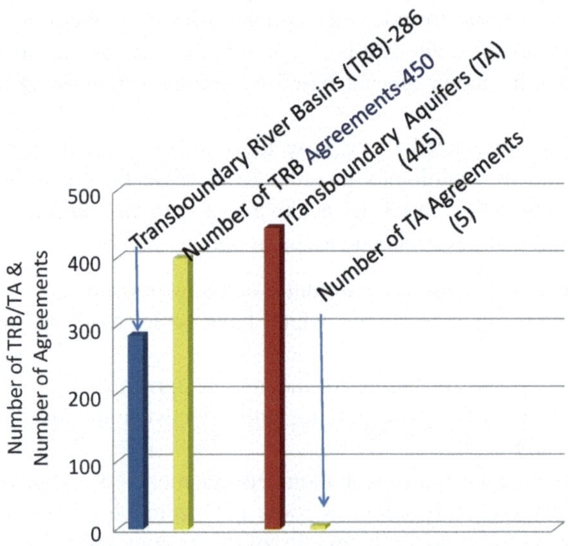

Fig. 6.10 Number of TRB/TA and number of agreements (Oregon State University 2018)

Table 6.1 Number of international agreements on transboundary aquifers

Continent/year/name/countries	Remarque/map
Europe • 1977/2007 Genevese aquifer Swiss and France	Local authorities
Africa • 1992/2000 I/2000 II Nubian Sandstone A.S. Egypt, Lybia, Chad, Sudan • 2002 Northwestern Sahara A.S. Algeria, Lybia, Tunisia • 2009 Iullemeden A.S. Niger, Nigeria, Mali	
S. America • 2010 Guaraní A.S. Argentina, Brazil, Paraguay, Uruguay	

6.5.4 *Negotiation Tools*

Achieving the establishment of proficient and healthy negotiations and facilitating the reaching of an agreement is only one part of any water diplomat's responsibilities. One of the chief aims of modern water diplomacy is to bear in mind the need for sustainable agreements. It appears that, actually, agreements only achieve a state of sustainability once they are converted into a strong and stable overhead structure, sometimes with supranational powers, that helps prolong an agreement's life expectancy and ensure that its governing principles remain guaranteed, notably despite changing management teams over the years.

A number of psychological factors, emanating from the way negotiations were held, impact the sustainability potential in any agreement, as we will see, but it is especially "institutionalisation" which plays a central role in prolonging agreements life expectancy. The eminence of institution-building, both in effective transboundary water management and in preventive water diplomacy, cannot be over-emphasised and twentieth and early twenty-first century water management offers groundbreaking lessons for the conception and implementation of modern international transboundary water institutions.

The following Fig. 6.11 illustrates the road-map for a "getting to yes" negotiation (Fisher and Ury 1981).

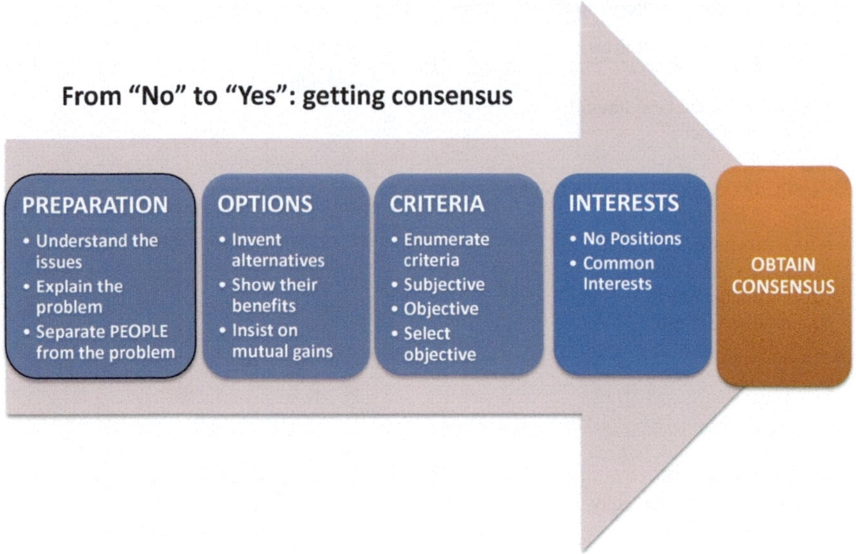

Fig. 6.11 The road-map from "no" to "yes" (based on Fisher and Ury 1981)

The various stages through which a negotiation can reach consensus and furthermore an institutionalised state, may be described as follows (Fig. 6.11):

1. PREPARATION
2. DEVELOPMENT OF ALTERNATIVE OPTIONS
3. AGREE ON EVALUATION OBJECTIVE CRITERIA
4. ANALYSE OPTIONS BASED ON CRITERIA AND COMMON INTERESTS
5. GET A WIN-WIN CONSENSUS

In particular, we will further analyse how, through building River Basin Organisations (RBOs), the ultimate goal of any attempt at jointly managing water through cooperation can be achieved. Finally, in Part III of the book, and particularly in Chaps. 7 and 8, examples of functional transnational basin management organisations will be presented to further enhance the link between theory and practice.

Keys to Achieve Sustainable Agreements To achieve a durable settlement and ensure that agreements be long-lived, there are at least three types of interests that generally must be met (UNITAR 2017). These interests are:

- Substantive–Content needs, money, time, goods, or resources.
- Procedural–Needs for specific types of behaviour or the "way that something is done".
- Psychological or relationship—Needs that refer to how one feels, how one is treated, or conditions for ongoing relationships.

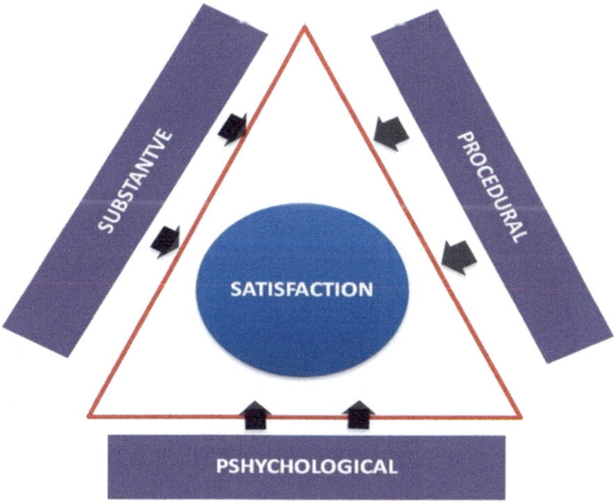

Fig. 6.12 The satisfaction triangle

These can be seen graphically on three sides in the satisfaction triangle (Fig. 6.12). Ideally, any conflict management process would be designed to seek an optimal satisfaction of the procedural, psychological, and substantive interests of each of the parties. Frequently, through designing conflict management and public involvement processes, professionals implicitly or subconsciously behave by successive partial satisfaction.

We know we have achieved "procedural satisfaction" when the parties agree to use the same approach again. "Substantive satisfaction" is done when negotiators think that water resources management objectives has been reached. "Psychological satisfaction" is personal and more complex to conceive. We may also have a situation that is high on the substantive or content aspects of the conflict but relatively low on the psychological and procedural aspects. Public involvement and public awareness require a maximisation of all three concerns, i.e. procedural, substantive and psychological. This is often very difficult and, in fact, often beyond the skills of many water resources professionals. The following Table 6.2 compares attitudes based on interests and positions (adapted from UNITAR 2017).

The table contains two columns and the words under each column indicate how people may feel when they perceive they have either "won" or "lost" in a dispute. Now, the following questions can be posed: "What possibility exists for a durable settlement if one party feels the way that is described by the words in column 1 and the other party feels the way described by the words in column 2?" Can a durable settlement exist when both parties feel as described by the words expressed in column 2?"

The answer in both cases is that there is little or no possibility. Parties must come close to feeling as described by the words in column 1 for durable settlements to exist. The point for diplomats with a concern for water-related issues and other professionals with expertise in water resources is that all must explicitly reach out to design processes that will result in such feelings as a condition to sustainable agreements.

Table 6.2 The win/loss satisfaction procedures

Interest-based "soft" negotiators	Position-based "hard" bargainers
• Resources are not limited • Interests can be addressed—reached • Focus on interests, not positions • Look for objective and fair standards • Search multiple satisfactory solutions • Find cooperative win-win solutions • People and issues are separate; respect people, bearing hard on interests	• Resources are limited • Adverse behaviour • Win—loss solutions • The goal is to win as much as possible • Concessions are a sign of weakness • There is a right solution—mine • Be on the offensive at all times

References

6th World Water Forum—WWF (2011) INWEB: TSG 1.5.2 coordinator. Marseille, France. http://www.worldwaterforum6.org/home/. Accessed 23 Dec 2017

Armitage D, de Loë RC, Morris M, Edwards TWD, Gerlak AK, Hall RI, Huitema D, Ison R, Livingstone D, MacDonald G, Mirumachi N, Plummer R, Wolfe BB (2015) Science-policy processes for transboundary water governance. Ambio. Royal Swedish Academy of Sciences, Springer. https://doi.org/10.1007/s13280-015-0644-x

Del Castillo L (2016) An illustration of transboundary groundwater management and protection legal aspects-the law of transboundary aquifers: scope and rippling effects. In: Fried J, Ganoulis J (eds) Transboundary groundwater resources-sustainable management and conflict resolution. Lambert Academic Publishing, Saarbrücken, pp 250–259

Delli PJ, Wolf AT (2009) Managing and transforming water conflicts. Cambridge University Press, New York, p 384. https://doi.org/10.1017/CBO9780511551536

Dombrowsky I (2009) Revisiting the potential for benefit sharing in the management of transboundary rivers. Water Policy 11(1):25–140

European Commission (2016) Water information system for Europe. Introduction to the new EU Water Framework Directive. http://ec.europa.eu/environment/water/water-framework/info/intro_en.htm

Fisher R, Ury W (1981) Getting to yes: negotiating agreement without giving in. Penguin, New York

Fried J, Ganoulis J (eds) (2016) Transboundary groundwater resources-sustainable management and conflict resolution. Lambert Academic Publishing, Saarbrücken, p 259

Knieper C, Pahl-Wostl C (2016) A comparative analysis of water governance, water management, and environmental performance in River Basins. Springer Science+Business Media, Dordrecht. Water Resources Management. https://doi.org/10.1007/s11269-016-1276-z

Mianabadi H, Moster E, van de Giesen N (2015) Transboundary river basin management: factors influencing the success or failure of international agreements, chap 7. In: Hipel KW et al (eds) Conflict resolution in water resources and environmental management. Springer International Publishing, Cham. https://doi.org/10.1007/978-3-319-14215-9_7

Nicol A, Mtisi S (2003) Politics and water policy: a Southern Africa example. IDS Bull 34(3). https://doi.org/10.1111/j.1759-5436.2003.tb00076.x

Oregon State University (2018) http://transboundarywaters.science.oregonstate.edu/

Phillips DJH, Allan JA, Claassen M, Granit J, Jägerskog A, Kistin E, Patrick M, Turton A (2008) The transcend-TB3 project: a methodology for the transboundary waters opportunity analysis (the TWO analysis). Prepared for the Ministry of Foreign Affairs, Sweden. http://www.siwi.org/sa/node.asp?node=168

Qaddumi H (2008) Practical approaches to transboundary water benefit sharing. ODI Working Paper 292

Quevauviller P (2016) The EU groundwater regulatory framework and its application to transboundary groundwater bodies. In: Fried J, Ganoulis J (eds) Transboundary groundwater resources-sustainable management and conflict resolution. Lambert Academic Publishing, Saarbrücken, pp 147–171

Ramoeli P (2010) Sharing benefits from transboundary water management-results from workshop on experiences from the Mekong River, Zambezi and the Nile, Vientiane, Lao. SADC

Referenced in the SADC Paper

SADC (2010) Concept paper prepared for the 4th SADC RBO Workshop (2010) on benefit sharing in the context of transboundary water resources management and development (Southern African Development Community). http://www.orangesenqurak.org/UserFiles/File/SADC/SADC%20concept%20paper_benefit%20sharing.pdf

Sadoff CW, Grey D (2002) Beyond the river: the benefits of cooperation on international rivers. Water Policy 4:389–403

Sadoff CW, Grey D (2005) Cooperation on international rivers. A continuum for securing and sharing benefits. Water Int 30(4):1–8

Sadoff CW et al (2008) Share: managing water across boundaries. IUCN, Gland

Schmeier S (2010) Governing international watercourses-perspectives from different disciplines: a comprehensive literature review. Hertie School of Governance-Working Papers No 53

Soliev I, Wegerich K, Kazbekov J (2015) The costs of benefit sharing: historical and institutional analysis of shared water development in the Ferghana Valley, the Syr Darya Basin. Water 7:2728–2752

Stephan RM (2016) International groundwater law and the UNILC. In: Fried J, Ganoulis J (eds) Transboundary groundwater resources-sustainable management and conflict resolution. Lambert Academic Publishing, Saarbrücken, pp 73–84

Tafesse T (2001) An appraisal of transboundary water dispute resolution in the Nile Basin. http://www.inweb.gr/twm4/abs/TAFESSE%20KASSE%20%20Tesfaye.pdf

Tanzi A (2014) Comparing the 1992 Helsinki water convention with the 1997 UN New York convention on international watercourse: harmonization over conflict. Quest Int Law 8:19–33

Turton A (2008) A South African perspective on a possible benefit-sharing approach for transboundary waters in the SADC region. Water Alternat 1(2):180–200. www.water-alternatives.org

UN Watercourses Convention Online User's Guide. http://www.unwatercoursesconvention.org/

UNDESA UN Water International Decade for Action (2014) 'Water for Life' 2005–2015. http://www.un.org/waterforlifedecade/transboundary_waters.shtml

UNITAR (2017) UN Institute for Training and Research. Introduction to water diplomacy, lecture notes, module II. https://www.learnatunitar.org/. Assessed 26 Dec 2017

UN-Water Status Report on IWRM and Water Efficiency Plans for CSD-16 (2008)

UN-Water thematic paper Transboundary Waters: sharing benefits, sharing responsibilities (2008)

USAID-IUCN Ecosystem approaches and integrated water resources management-interrelated approaches Fact Sheet N°1

Wolf A (1999) Criteria for equitable allocations: the heart of international water conflict. Nat Resour Forum 23:3–30

Wolf A (2010) Sharing water, sharing benefits: working towards effective trasnboundary water resources management. UNESCO PC-CP, IHP Division of Water Sciences

Part III
Transboundary Hydro-Governance in Practice

Chapter 7
From Theory to Practice: Case Studies Worldwide

We forget that the water cycle and the life cycle are one.
Jacques Yves Cousteau, Undersea Explorer, Filmmaker

Abstract To illustrate what is hydro-governance in practice and its diversity, but to try to identify features that could serve as a basis for common hydro-governance guidelines for both surface waters and groundwater, either separately or jointly managed, we propose several examples of transboundary hydro-governance for both river basins systems and groundwater systems, where most of them introduce and describe International River Basin Organization, called IRBO's or TBO's (Transboundary Basin Organizations), with some exceptions we present as well.

Keywords River basin · Practice · Case histories · IRBO's · TBO's

7.1 Global River Basins Organisations

7.1.1 The International Commission for the Protection of the Danube River (ICPDR)

The Danube River Basin is considered as the most international river basin in the world with 19 riparian countries, covering approximately 801,463 km^2, the and extending from the origination of the Danube River in Germany to the Romanian and Ukrainian shores along the Danube Delta and the Black Sea (Fig. 7.1).

The European Commission of the Danube is probably one of the oldest IRBOs worldwide as it was set by the Treaty of Paris of 1856, for the administration of the Danube River, primarily to ensure free navigation along the river for all European countries. Besides navigation, other issues, such as water quality, gained importance and, in 1985, the riparian countries signed the "Declaration of the Danube Countries to Cooperate on Questions Concerning the Water Management of the Danube (Bucharest Declaration)". This document committed the countries to an integrated approach in water management, on the principle that the environmental quality of the river depends

J. Ganoulis, J. Fried, *Transboundary Hydro-Governance*,
https://doi.org/10.1007/978-3-319-78625-4_7

Fig. 7.1 The Danube River basin

on the environment of the basin, and a basin-wide unified monitoring network was then established. In 1991, an Environmental Programme for the Danube River Basin (EPDRB) was initiated, stating that each riparian country would:

- adopt the same monitoring systems of assessing environmental impact;
- address the issue of liability for cross-border pollution;
- define rules for the protection of wetland habitats,
- define guidelines for development to conserve areas of ecological importance or aesthetic value.

A major task of the EPDRB has been to develop a Strategic Action Plan (SAP) concerned with measures as concrete as possible and short term goals, with realistic costs, while the riparian countries were developing the Convention on Cooperation for the Protection and Sustainable Use of the River Danube or Danube River Protection Convention (DRPC), which was signed in 1994 and constitutes the overall legal instrument for cooperation and transboundary water management in the Danube River Basin, with the main objective of ensuring that the surface waters and groundwater within the Danube River Basin are managed and used on a sustainable and equitable basis. To implement the Convention and the SAP, and ensure these objectives are accomplished, the International Commission for the Protection of the Danube River (ICPDR) has been established, comprised of the delegations of the contracting parties to the Convention but open to other organizations, and immediately entrusted with the implementation of the SAP.

Composed of national delegates, at high ministerial level, technical experts, representatives of the Civil Society and scientists, the ICPDR is comprised of an Ordinary Meeting Group, taking the political decisions, a Standing Working Group, providing political guidance and Technical Expert Groups and Task Groups, preparing the technical background documents. It is supported by a Secretariat, which maintains DANUBIS, the ICPDR information system. A quite significant priority objective of the ICPDR is the sharing, exchange and harmonization of all data concerning the Danube River, such as water quantity and quality, and the coordination of warning and alarm systems and emergency plans. In particular, the establishment of uniform standards of data collection and exchange has been a prime focus of the contracting parties. There are also joint data collection and survey efforts, and a technical body, the Information Management and Geographical Information System Expert Group, charged with maintaining the overall data information system.

The ICPDR relies on eight Expert Groups formed by national experts from the contracting countries and delegates from observer organizations, which provide technical support to the ICPDR and are completed by four Task Groups dealing with specific issues. For instance, communication and public participation being key issues, a Public Participation Expert Group supports ICPDR activities in this domain, including public consultation measures for the Danube River Basin Management and Flood Risk Management Plans, publications and outreach initiatives such as educational programs for the general public including an International Danube Day and the creation of a Danube Box, which is a teaching toolkit to "give local schoolchildren a greater understanding of the river, the threats posed to the river, and the need to preserve water resources" (International Waters Governance 2011).

Besides being an operative instrument in charge of implementing the Convention's objectives and provisions, the ICPDR is also an executive tool as it has the legal capacity "to contract, to acquire and dispose of immovable and movable property, to institute or respond to legal proceedings and to take such other action as maybe necessary or useful for its purposes and activities". This legal capacity is determined by the law of the country where its secretariat is located, Austria, and the ICPDR is represented by its President.

Implementing the Convention's objectives and provisions means protecting the Danube Basin's water resources for future generations by preserving the natural balance of those waters, addressing risks from toxic chemicals, preventing the environmental and ecological damages caused by floods, and maintaining the health and sustainability of the region's river systems. For such purpose, the ICPDR has initiated or supported many basin-wide projects, such as, among others, the Danube Pollution Reduction Programme (1997–1999), the Danube Regional Project (2002–2006) for "Strengthening the Implementation Capacities for Nutrient Reduction and Transboundary Cooperation in the Danube River Basin", or educational programs to educate the general public about the threats to the Danube River Basin and how to get involved.

The ICPDR has strong relationships with both UNDP-GEF and the European Commission, and also both formal and informal relationships with other international organizations and corporate partners. Funding does come from the contracting parties in equal parts, which are also expected to pay for their own costs of participation and the costs of monitoring and assessment undertaken in their territories. Of course, particular projects will be funded by the UNDP-GEF, the European Union, the World Bank and other public and private partners. Finally, the ICPDR uses multiple public and private stakeholders to determine policy priorities and implement specific programs.

7.1.2 The International Commission for the Protection of the Rhine River (ICPR)

The International Commission for the Protection of the Rhine River (ICPR) is operative since 1964, although it was informal since 1950. It implements the Convention on the Protection of the Rhine, signed in 1999, and entered into force in 2003. Among other objectives, the preservation, improvement and sustainable development of the Rhine ecosystem are central elements of the convention. This target was fixed against the background that the Rhine is an important European navigation lane (Fig. 7.2) and is supposed to continue to serve different uses, such as drinking water production or hydro-power. Keeping in mind preservation and improvement of the North Sea, the restoration of the Rhine has an additional international dimension. The Convention integrates surface water and groundwater as it applies to both the Rhine and the groundwater interacting with the Rhine. Its goal is, through the use of a comprehensive approach, to increase multilateral cooperation in the sustainable development of the Rhine's ecosystem, by, for instance and among others, maintaining and improving the quality of the Rhine's waters (including groundwater) through the prevention, reduction, or elimination of pollution caused by noxious substances, nutrients from point sources (such as industries and municipalities), diffuse sources (such as agriculture and traffic) and shipping; ensuring the environmentally sound management of water resources; and producing drinking water from the Rhine.

Consisting of the delegations of the contracting countries, possibly assisted by experts, the Commission is comprised of the Plenary Assembly, which meets once a year and issues recommendations to the countries, while Conferences of Rhine Ministers decide on important political issues. Technical questions are dealt with in working and expert or project groups with permanent or fixed-term mandates, and passed on to the Strategy Group, made up of the heads of delegations, preparing the Plenary Assembly and assisted by the Secretariat. Technical questions for instance comprise water quality, groundwater, floods, warning and alarm procedures. States, Intergovernmental Organizations, Non-Governmental Organizations may be recognized as observers to the Commission. The Commission has legal personality, according to the legislation of the concerned contracting country. Decisions are taken at unanimity.

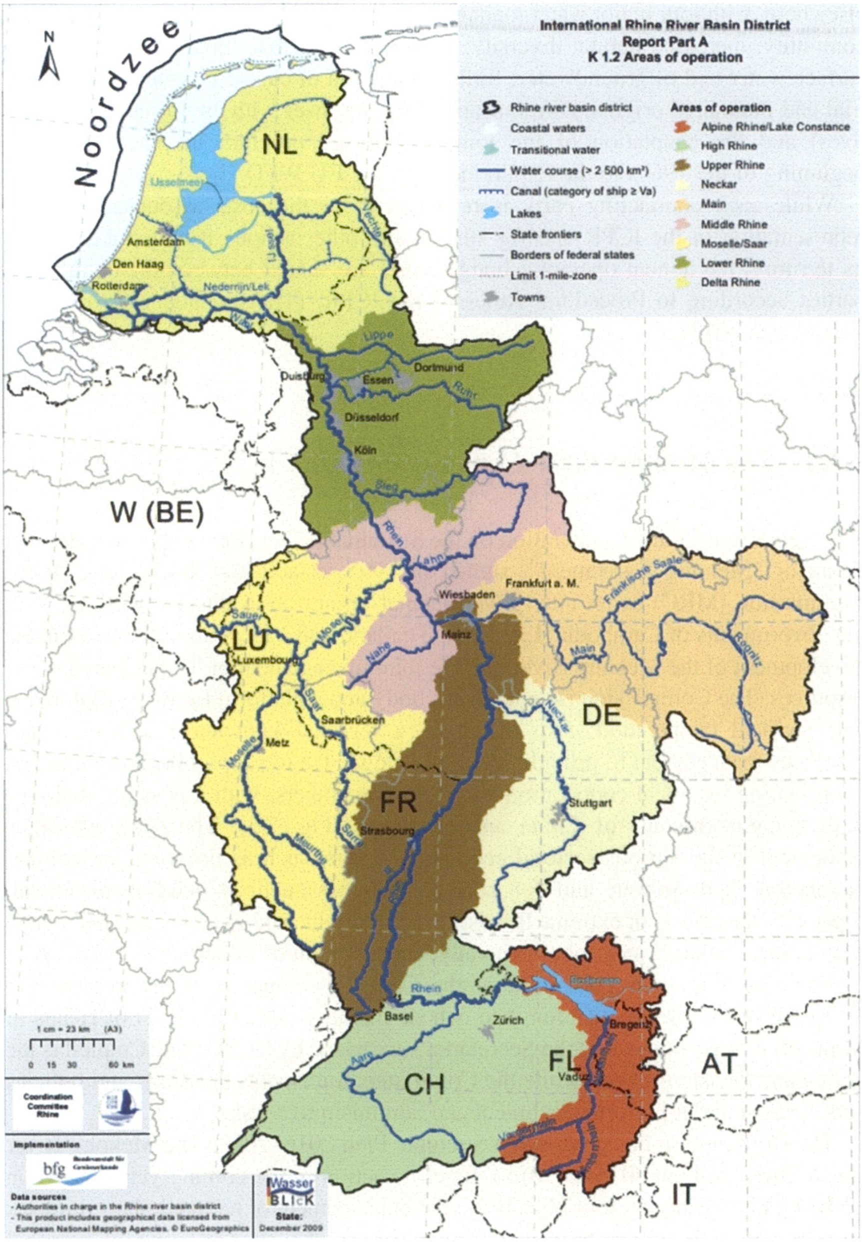

Fig. 7.2 The Rhine River basin

Many bilateral/multilateral agreements have been in place besides the Convention for many years, some dating from the beginning of the nineteenth century. The Commission has also developed an international River Basin Management Plan, as prescribed by the EU Water Framework Directive (EU WFD) we have presented elsewhere, with four major water management issues (Restoration of biological river continuity, increased habitat diversity; reduction of diffuse inputs interfering with surface water and of groundwater; further reduction of classical pollution of industrial and municipal origins; harmonisation of water uses with environmental objectives) and the adaptation of the joint monitoring program in place since the beginning of the 1960s to the requirements of the EU WFD.

While each contracting party is responsible for the costs associated with its representation in the ICPR and for studies and other actions it undertakes within its territory, the annual operating budget costs are shared between the contracting parties according to Procedural Rules of the ICPR. (Transboundary Cooperation Fact Sheets 2012)

7.1.3 The Mekong River Commission (MRC)

The Agreement for the Cooperation on the Sustainable Development of the Mekong Basin or “Mekong Agreement”, signed in 1995, established the Mekong River Commission (MRC) as an inter-governmental organization directly working with the governments of Cambodia, Lao PDR, Thailand and Viet Nam in the sustainable development of the Mekong River and the joint management of its the shared water resources. The Commission is an advisory body and a platform for water diplomacy and regional cooperation, also acting as a regional knowledge hub on water resources management to provide scientific knowledge to support decision-making, coordinating technical cooperation between its members, with a possible dialogue with the governments of China and Myanmar (Fig. 7.3). The Commission is competent in the various sectoral economic sectors, such as fisheries, agriculture, navigation, hydropower, and for ecological issues such as flood management, especially the effects of extreme floods, severe droughts and sea level rise associated with climate change, and, more generally, conservation of important ecosystems.

The Commission is comprised of the Council, composed of the ministers in charge of water, the Joint Committee, whose members are at the level of Heads of Ministerial Department, and the Secretariat supervised by a CEO. The Council is the policy and decision-maker, while the Committee implements decisions and policies and the Secretariat performs technical and administrative tasks.

The Commission has produced a Strategic Plan 2016–2020 (The Mekong River Basin Strategic Plan 2016–2020) to direct its action in the coming years based on IWRM and focusing on four objectives: the enhancement of national plans, projects and resources based on basin-wide perspectives; the strengthening of regional cooperation; better monitoring and communication of Basin conditions; and leaner River Basin Organization. The implementation of this Strategic Plan should, hopefully, answer the comments concerning the way the Commission is operating proposed by the Danish Institute for International Studies (DIIS) in its 2013 Report (Jensen and Lange, DIIS 2013).

Fig. 7.3 The Mekong River basin

A Forum of the Stakeholder has been held in June 2017 (Regional Stakeholder Forum 2017) and produced the Mekong Adaptation Strategy and Action Plan (MASAP) with the purpose of "setting out the strategic priorities and actions at basin level to address climate change risks on the basin and strengthen basin-wide resilience, contributing to ensuring sustainable development of the Mekong River Basin in line with the 1995 Agreement on the Cooperation for the Sustainable Development of the Mekong River Basin". It produced 7 strategic priorities, considered as the pillars of the MASAP:

1. Mainstreaming Climate Change into regional and national policies, programs and plans;
2. Regional and international cooperation and partnership on adaptation;
3. Implementation of transboundary, gender sensitive adaptation options;
4. Access to adaptation finance;
5. Monitoring, data collection and sharing;
6. Capacity development for climate change adaptation strategies and plans;
7. Outreach of MRC products on climate change and adaptation.

In the list of references, the web address of the Forum has been given and we recommend reading the MASAP part.

The DIIS Report mentioned above gives some interesting general comments about RBOs, which we find worthwhile to present hereafter as a quotation:

> The road to political relevance for the MRC and for RBOs in general is evident through more proactive engagement in the geopolitical dynamics of transboundary water governance. Conflicts over water developments cannot be avoided, but... if the RBO wants to retain its decision-making forum and achieve more political relevance as a conflict mediator, it needs to be able to present knowledge on development trade-offs, options for benefit-sharing and compromises over conflicts. These RBOs also play an important role in setting the regional agenda, and even though political realities are frequently very different from the ideals embodied by IWRM and others, these ideals are important in setting the right agenda and raising expectation levels of delivery and decision-making in multi-country river basins. While governments cannot be expected to negotiate issues of national importance publicly, more transparent water-related solutions may be possible... The MRC...specifically need to gear up (its) capacity to deal effectively with the evolving water governance challenges in the basin(s)... In their future governance efforts, the RBOs also need to address the polycentric realities, geopolitics and interests of individual governments, investors and civil-society stakeholders. Opposing interests may increasingly turn into tension and conflicts in both river basins as their water regimes close.
>
> Pertinent issues for the RBOs will include relevant knowledge production, the capacity to engage in development diplomacy and the conflict-resolution skills needed in political discussions over water developments. Also, RBOs need to engage more directly with civil-society groups engaged in knowledge production (e.g. WWF) or advocacy (e.g. TERRA, Justicia Ambiental, International Rivers), as well as private investors...While we suggest a reorientation of focus by the RBOs, we are also well-aware that the level of ambition of their multilateral water-governance mandates need to be carefully considered in the face of the already existing reality of polycentric governance (Kinna and Rieu-Clarke 2017).

7.1.4 *The Senegal River: "Organisation de Mise en Valeur du Fleuve Sénégal" (OMVS)*

To sustainably manage the River Senegal, shared by four countries, namely Mali, Mauritania, Guinea and Senegal (International Waters Governance 2018; Burchi et al. 2003) (Fig. 7.4), an International River Basin Organization (IRBO) has been established in 1972, the OMVS (Organisation de Mise en Valeur du fleuve Sénégal),

Fig. 7.4 The Senegal River basin

by Mali, Mauritania and Senegal, with Guinea as an observer. The legal and regulatory framework of the OMVS consists of two conventions, the Convention Concerning the Status of the Senegal River (Convention Relative au Statut du Fleuve Sénégal) ("Senegal River Convention"), and the Convention Establishing the Organization for the Development of the Senegal River (Convention Portant Création de l'Organisation pour la Mise en Valeur du Fleuve Sénégal or "OMVS Convention").

The Senegal River Charter has been signed in 2002, and its main objectives are to fix principles and modalities for the sharing out of the waters of the Senegal River among the different areas of use; to determine rules related to the preservation and to the protection of environment, particularly as regards fauna, flora, and ecosystems; and finally, to define the framework and the modalities of public participation in decision-making as far as water resources management of the river is concerned.

The resource is not allocated to riparian states in terms of volumes of water to be withdrawn, but rather to uses as a function of possibilities. The various uses can be for agriculture, which is the main activity, then livestock raising, inland fishing, fish farming, tree farming, fauna and flora, hydroelectric energy production, urban and rural drinking water supply, health, industry, navigation and the environment.

The interests of the member countries do not fully coincide: Mali wants to maintain river levels to obtain navigable access to the sea, although navigation is not a major objective, and energy produced by a common dam, the Manantali dam; Mauritania and Senegal are interested in power production and irrigation.

The OMVS has full legal capacity and the power to enter contracts, acquire and dispose of property, receive donations, subsidies, legacies and other gifts, request loans, apply for technical assistance, and institute legal proceedings.

The Council of Ministers ("Council") is the legal representative of the OMVS and can delegate the legal authority needed to exercise powers to the High Commissioner. The OMVS is charged with implementing the Senegal River Convention, promoting and coordinating development studies and works on the Senegal River Basin within the Member States and carrying out all technical and economic functions conferred to it by the Member States.

The Conference of Heads of State and Government ("Conference") is the chief decision-making body and is responsible for setting the general policies of the OMVS, while the Council, whose decisions taken unanimously are binding for the Member States, sets priorities and formulates the policies for managing the Senegal River, developing its resources, and promoting the cooperation of states around the Senegal River. The Council also approves the budget and determines the funding from the Member States.

The Office of the High Commissioner, executive body of the OMVS and appointed for 4 years renewable, implements the decisions of the Council of Ministers, is responsible for the administration and staff of the OMVS and exercises the powers delegated to it by the Council of Ministers. It also implements studies and projects relating to hydrology and agriculture, solicits funds for projects, and coordinates the development and exploitation of common works. The Permanent Water Commission allocates water rights among the Member States and different sectors, including industry, agriculture, and transport.

The Advisory Committee provides advice to the OMVS, while the Regional Planning Committee advises the OMVS on the regional development plans of the Member States and their impact on the basin's resources. The National Offices coordinate with the OMVS regarding project management and implementation, as well as for activities within the Member States.

Disputes are resolved by mediation, possibly submitted to the Commission of Mediation, Conciliation and Arbitration of the Organization of African Unity, whose decisions can be appealed to the International Court of Justice. Of course, all needed information should be provided to the OMVS and exchanged between the Member States. The Member States each contribute to the OMVS ordinary budget. The costs and expenses for common works are shared among the Member States in proportion to the benefits received by each Member State from the work. The sharing of costs for common works is to be reassessed periodically. The Convention on the Financing of Common Works provides a framework for methods of financing, such as contributions, loans, and subsidies.

An interesting aspect that could be of interest for IRBOs in general is the part played by Non-State Actors in the decision-making process. Three groups of Non-State Actors are involved in the decision-making process: users or local populations; territorial communities; and NGOs and decentralized committees of coordination. Altogether, public participation in decision-making is still far from realized in practice, although it is a significant theoretical component of hydro-governance (Sène et al. 2007).

7.1.5 *The US-Mexico International Boundary and Water Commission (IBWC)*

An International Boundary Commission was established in 1889 between the US and Mexico, and became the International Boundary and Water Commission (IBWC) by the treaty of 1944 on "the Utilization of Waters of the Colorado and Tijuana Rivers and of the Rio Grande", more specifically including water. It has responsibility for applying the boundary and water treaties between the United States and Mexico and settling differences that may arise in their application.

The IBWC is an international body composed of the United States Section and the Mexican Section, each headed by an Engineer-Commissioner appointed by his/her respective president. Each Section is administered independently of the other. The United States Section of the International Boundary and Water Commission (USIBWC) is a federal government agency and is headquartered in El Paso, Texas. The IBWC operates under the foreign policy guidance of the Department of State. The Mexican Section (CILA, Comision Internacional de Limites y Aguas) is under the administrative supervision of the Mexican Ministry of Foreign Affairs and is headquartered in Ciudad Juarez, Chihuahua, Mexico. Each Section maintains its own engineering staff, a secretary and such legal advisers and other assistants as it deems necessary and has field offices strategically located along the border to operate and maintain joint works and provide support in resolving binational issues. The Commissioners meet frequently. Each Government funds the cost of the operation of its Section of the IBWC, while the cost of joint projects is shared between the U.S. and Mexican Governments, in proportion to their respective benefits in cases of projects for mutual control and utilization of the waters of a boundary river, unless the Governments have predetermined by treaty the division of costs according to the nature of a project. In cases of man-made works in one country or operations in one country causing or threatening to cause damage in the other country, the cost is borne by the Government in whose territory the problem originated. The IBWC supervises the execution of the joint projects, each country doing its share as agreed.

The mission of the IBWC is to apply the rights and obligations which the Governments of the United States and Mexico assume under the numerous boundary and water treaties and related agreements, and to do so in a way that benefits the social and economic welfare of the peoples on the two sides of the boundary and improves relations between the two countries. Those rights and obligations include:

- distribution between the two countries of the waters of the Rio Grande and of the Colorado River;
- regulation and conservation of the waters of the Rio Grande for their use by the two countries by joint construction, operation and maintenance of international storage dams and reservoirs and plants for generating hydroelectric energy at the dams;
- regulation of the Colorado River waters allocated to Mexico;
- protection of lands along the river from floods by levee and floodway projects; solution of border sanitation and other border water quality problems;
- preservation of the Rio Grande and Colorado River as the international boundary; and demarcation of the land boundary.

Decisions of the Commission are subject to the approval of the two Governments and are recorded in the form of Minutes done in the English and Spanish Languages, signed by each Commissioner and attested by the Secretaries. Copies are forwarded to each Government within 3 days after being signed. Once approved by both Governments, the Minutes enter into force as binding obligations of the United States and Mexican Governments.

In addition to cooperative projects undertaken to implement existing treaties and international agreements, the U.S. and Mexican Commissioners make recommendations to their respective Governments for resolution of new or anticipated boundary or water problems. Early detection and evaluation of the problem and the development of measures for resolution are a part of the mission of the IBWC. Most problems are resolved by the development of new projects. The need for development of new cooperative projects may also be brought to the attention of the IBWC by one or both Governments, or by state or local authorities through their respective Section of the IBWC. If the findings of the IBWC joint investigations, recorded in a joint report of the Principal Engineers of the two Sections, show that a cooperative project is needed, is feasible and can be justified as an international project, the IBWC may endorse the findings in a Minute and recommend the project to the two Governments.

The case of groundwater is rather interesting: though groundwater has been neglected in the 1944 treaty, the IBWC has dealt successfully with groundwater problems on a pragmatic, ad hoc basis (Utton 1978) and has had a partial mandate to explore mechanisms of binational cooperation on transboundary groundwater since the Salinity Crisis drew attention to the problem in the late 1960s. Minute 242, signed in 1973, charged the Commission with monitoring groundwater withdrawals in the San Luis Rio Colorado—Yuma area and required the two governments to inform each other of any new developments that might affect the other party (IBWC 1973). It went even further, and contemplated the need for a comprehensive agreement on border groundwater (Mumme 2001). In 1998, IBWC, US Section, and CILA have published a groundwater database concerning the transboundary aquifers of the City of El Paso/Ciudad Juarez Area (Hueco-Tularosa aquifer, Southeastern Hueco aquifer, Rio Grande aquifer), which integrates official groundwater data from the US and Mexico into one database. And the Binational Study of the transboundary San Pedro Aquifer (see Sect. 7.2.4 hereafter) was published, as a water quality report, in 2016 (Binational Study 2016; Megdal 2017).

7.1.6 The US-Canada International Joint Commission (IJC)

Article VII of the Boundary Waters Treaty, signed in 1909 by the USA and Canada, creates "an International Joint Commission of the United States and Canada, composed of six commissioners, three on the part of the United States appointed by the President thereof, and three on the part of the United Kingdom appointed by his Majesty on the recommendation of the Governor in Council of the Dominion of Canada." The IJC has two main responsibilities: regulating shared water uses and investigating transboundary issues and recommending solutions. It is not limited to water but also deals with the protection of the general transboundary environment, e.g. the improvement of transboundary air quality International Joint Commission (IJC) (2017a, b).

The treaty provides general principles, rather than detailed prescriptions, for preventing and resolving disputes over waters shared between the two countries and for settling other transboundary issues. The specific application of these principles is decided on a case-by-case basis.

The Commissioners are assisted by three types of staff, a Canadian section staff, a US section staff and a Great Lakes Regional Office staff. Besides, the IJC establishes boards to assist in carrying out activities under references, orders of approval, treaties and agreements. It may also establish task forces and work groups to support specific activities.

To demonstrate how the Commission will work closely with its boards and local citizens in shared waters to assist governments' efforts in the prevention and resolution of binational disputes, a Strategic Plan has been developed for 2015–2020, based on the vision of healthy shared waters for present and future generations. From a water governance point of view, the Plan provides a modern perspective, whose five main priorities are worth mentioning hereafter:

The Commission will continue to fulfil its core responsibilities using the best science, a watershed approach, adaptive management and local binational partnerships. A significant responsibility of the Commission is to develop innovative scientific tools and approaches to support sound water management. The Commission has scientific and technical responsibilities that include the regulation of water levels and flows, reporting on the achievement of water quality objectives and developing tools to help resolve issues at the watershed level in certain waters along the international boundary.

The Commission's role is also to alert governments about emerging issues that could become problematic between the two countries in the coming years and the Plan has identified three such issues, namely over-enrichment by nutrients related to phosphorus, pollution by heavy metals, and climate change impacts to water availability and water quality.

The Commission should fulfil its role and responsibilities under the Great Lakes Water Quality Agreement (GLWQA) such as the provision of scientific and policy advice and recommendations to the governments on the water quality of the Great Lakes and engaging the public on water quality issues.

The Commission should sustain and enhance the effectiveness of boards guided by the Boundary Water Treaty, the GLWQA and other mandates, and direct its volunteer boards and task forces to conduct studies and report to the Commission on the conditions and operations of specific projects. The boards will be encouraged to strengthen binational relationships with various federal, provincial, state and municipal agencies, as well as with First Nations, Métis, Tribes and local stakeholders. Consistent with the modern approach to water governance, the Commission will further integrate water quantity and quality considerations, as appropriate, through the International Watersheds Initiative.

Finally, the integrative and coordinated ways of the Commission should be improved to increase the efficiency, effectiveness and accountability of its operations along and across the boundary, harmonizing the operations between the Commission offices on both sides of the border.

Notice: Introduced in 1997, the International Watersheds Initiative (IWI) is an approach to resolving transboundary water issues grounded on the belief that local communities, given appropriate assistance, are best placed to achieve solutions. This approach operates on an ecosystem focus, recognizing that ecosystems function as whole entities and should be managed as such, rather than being bound by traditional political boundaries. To make this approach into a reality, International Joint Commission (IJC) boards in watersheds along the Canada-U.S. border carry out IWI projects to help manage resources, promote communication, and conduct scientific studies. IWI has proved successful by helping inform, engage, and provide tools for decision makers at all levels to better address a broad range of contentious water-related issues along the border.

Groundwater has some specific characteristics. It was not mentioned in the 1909 Boundary Waters Treaty, and it is in 2000 only that it was explicitly considered, because of the need to protect groundwater recharge areas. As rather usual, groundwater has been neglected politically and scientifically, but this situation is slowly changing and groundwater seems to be gaining more attention in transboundary water issues: for instance, the IJC issued a comprehensive report on groundwater in the Great Lakes region in 2010 and the Great Lakes Charter Annex and accompanying set of agreements between two Canadian provinces and eight American states addresses groundwater extraction through its general prohibition on large-scale diversions from the Great Lakes Basin. In most cases, transboundary Canada-US water disputes are resolved through cooperative mechanisms and information sharing through action bodies such as the Abbotsford–Sumas International Aquifer Task Force, the Great Lakes Council of Governors, and the extensive bi-national cooperative framework of the IJC (Rivera 2015). Yet progress is slow and we will quote Rivera (2015) as a conclusion: "Many challenges remain, and many actions are needed: Legal and institutional instruments; shared management practices; social participation; innovative Science and Technique to assess Transboundary Aquifer Systems; policy based on informed decisions; open communication and education. A more participatory approach would be required to trigger cooperation between the two countries and make available groundwater-related information locked at national and local levels. A systematic approach of data collection to establish cohesive and accessible databases in order to allow global comparison and discovering of data/information gaps would be a first step".

7.2 Shared Aquifer Systems

7.2.1 The Geneva Aquifer Between France and Switzerland

The Transboundary Aquifer of the Geneva Region, shared between Switzerland and France, has been successfully jointly managed for more than 30 years by the State of Geneva and French Border Communities. It is probably the only successful practical case of transboundary groundwater governance and management and, therefore, we have decided to include it, using the description and analysis of de los Cobos (de los Cobos 2016). Yet, it must be stressed that it is not a cooperation between two countries but a more local cooperation between administrative entities of each country, namely the "Canton de Genève" for Switzerland, and the "département of Haute-Savoie" and its "prefecture" on the French side. Besides, there are not significant discrepancies between these entities as both regional entities benefit from the same language and culture and almost the same socio-economic level. From a hydrological perspective, enough surface water is available, with the Arve River, and the underlying aquifer is rather easily recharged due to favourable transmissivity and porosity.

Geneva's drinking water is provided by the Lake of Geneva (roughly 80% of total supply) and the Geneva transboundary aquifer, which straddles the canton of Geneva in Switzerland and the French department of Haute-Savoie (about 20% of total supply) with ten wells on the Swiss side and five wells on the French side (Fig. 7.5). Roughly 90% of the Genevese groundwater is in the canton of Geneva and 10% in France. In the 1970s water supply in Geneva was provided by two water companies (Société des Eaux de l'Arve, and Services Industriels de Genève—SIG), whereas on the French side several wells were exploited by various communities or syndicates.

During the 1960s and 1970s, because of uncontrolled over-pumping and the lack of coordination among distributing and beneficiary entities, groundwater levels fell drastically, to the point that certain dried out wells had to be closed. The decision was then taken to set up an artificial aquifer recharge plant to recover the use of the wells and store more water in the aquifer. An artificial recharge system was, therefore, inaugurated in 1980 to abstract water from the Arve River—which is the aquifer's main natural recharge source—treat it and channel it into the aquifer. This operation, effected via drains that are laid underground above the aquifer, ensured the maintenance of high groundwater levels as well as seasonal stockpiling of the drinking water resource. The system has been described in previous articles on artificial recharge (de los Cobos 2002, 2007).

In parallel to experimenting and setting the artificial recharge plant, a political will to develop a cross border project emerged naturally. Several meetings and discussions were held and agreements were drafted, to impose restrictions on use for as long as the diminishing groundwater, threatened with total depletion, was not recharged artificially. They would establish equitable cost sharing once the recharge plant was completed and operational and would then be of mutual benefit to users on

Fig. 7.5 The Genevese aquifer

both sides of the border. From the time the first Franco-Swiss meetings were held in 1972, it was noticed that groundwater resources were dramatically shrinking. The problem was not only affecting Geneva but the entire adjoining French region as well (Fig. 7.6).

Between 1973 and 1974, efforts focused on drawing up an inventory of drinking water resources in the region as well as on hydrogeological issues, such as losses, natural recharge, future pumping and fees, and the whole range of factors that needed to be considered in laying the bases of a future Franco-Swiss agreement. In 1975, the French declared that they would no longer use Genevese groundwater and would be turning to other French resources but wished to retain the possibility of later participating in and benefiting from artificial recharge. This decision certainly had an impact on cost coverage because, although SIG pursued their minimum extraction policy, the financial burden fell onto the shoulders of Société des Eaux de l'Arve which pumped exclusively Genevese groundwater. It was envisaged that the cost of

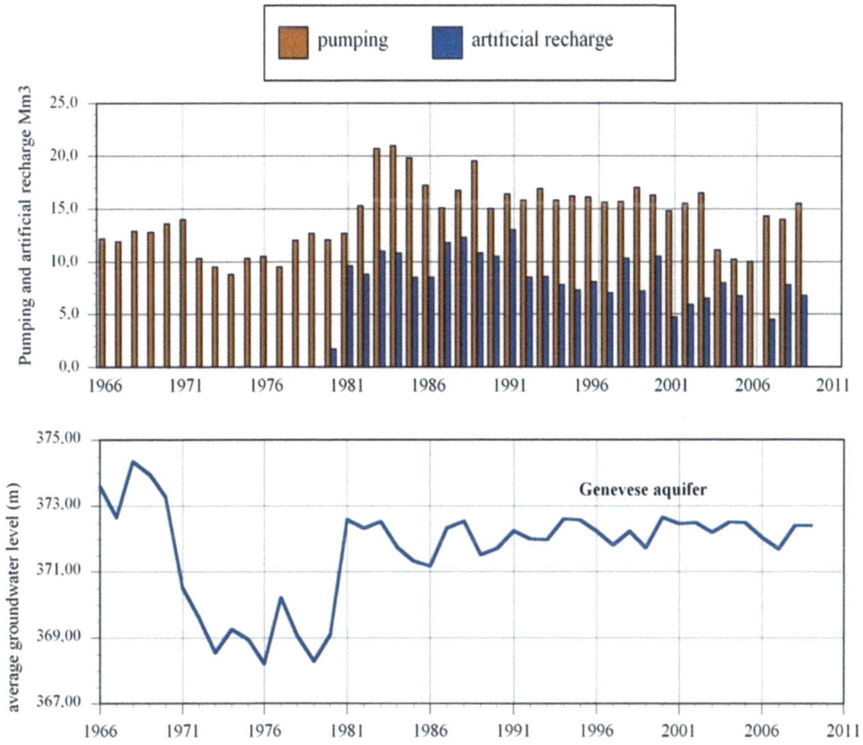

Fig. 7.6 Impact of the artificial recharge on the Genevese aquifer

artificial recharge (depreciation, interest, usage and renovation costs) would be spread among all groundwater users, irrespective of the origin of the water, whether natural or artificial recharged. The artificial recharge fee per cubic meter of water extracted could be less than 0.07 Swiss francs when the annual volume of water pumped was more than 20 million m^3. However, the fee could rise to 0.12 Swiss francs or more if the volume of water pumped was less than 10 million m^3.

As of 1977, artificial recharge benefits for users were decided as follows:

Société des Eaux de l'Arve: could continue to expand its network without having to seek out other resources (purchase of water from SIG or construction of a water pumping station to extract water from the lake).

SIG: could exploit all the existing pumping wells (100,000 m^3/day) and not have to build a new plant for treating lake water (100 million Swiss francs) which would only be used for a few weeks in the year.

French communities: If necessary, the French communities could extract supplementary amounts of water and pay their part of the costs beyond the annual minimum allowance of 2 million m^3 which corresponded to the volumes they extracted before artificial recharge was introduced.

All users: reserves of 15 million m^3 of drinking water would be stored in the best possible conditions (underground) to be constantly available in the case of acute shortage in one of their other plants.

Finally, an arrangement between the State of Geneva and Haute-Savoie in France was signed on 19 June 1978 for 30 years. The French authorities obtained an annual allowance of 2 million m^3. Once this quota was surpassed, the price per m^3 would be calculated based on an equation based on the costs of operating the plant, covered by SIG, depreciation, total pumping (Swiss and French), share in natural recharge and volume pumped by the French authorities minus the quota.

Following approval of the arrangement between the canton of Geneva and the prefecture of Haute Savoie, an advisory committee was set up to oversee the management of the recharge of the Genevese aquifer, which would also be responsible for dealing with problems relating to the use, maintenance and monitoring of the system as a whole. The committee would regularly assess the status of the resource in terms of pumping and artificial recharge. Stocked volumes of water (the amount of water each user was expected to use for the coming year) were discussed and accepted based on conditions relating to water quantity and quality and the operation of the artificial recharge plant.

Initially, on the Swiss side, the committee was comprised of two State Council designated representatives, a representative from the cantonal committee for water conservation and a representative of each water company (SIG and Société des Eaux de l'Arve). Experts could accompany committee members. One of the State Council designated representatives would chair the committee. The committee would be mainly responsible for reviewing the annual artificial recharge program (based on groundwater levels and forecasts of water user needs), the budget that was prepared in line with the recharge program, the operating costs and the results of water quality control tests. Provisions governing quantities reserved for each party for the coming year, water quality and the warning system in the event of accidental pollution were also to be discussed and adopted by the committee. The committee would meet once or twice year, alternatively in France and in Switzerland. It could also meet more frequently, if necessary. In 1988, the water company Société des Eaux de l'Arve was bought out by SIG, as of which point, Geneva had one single water supplier.

Between 2007 and 2008, the committee found itself facing new challenges, namely the extension of the 1978 arrangement after almost 30 years of operation, and, on the Swiss side, the transfer of Geneva State assets to SIG and, thus, the sale of the recharge plant to SIG. In view of these important events, a sub-working group of the committee was formed and met several times during 2007. This Franco-Swiss working group, having the necessary legal support, produced a basic document outlining the management of the Geneva shared aquifer for the next 30 years. It was to be ensured that the technical specificities agreed on by the parties would be placed within a global context that was recognized at administrative and political levels.

The first steps consisted of setting up the cross border legal bases for the new agreement. The 1978 arrangement did not have a legal premise that was grounded in

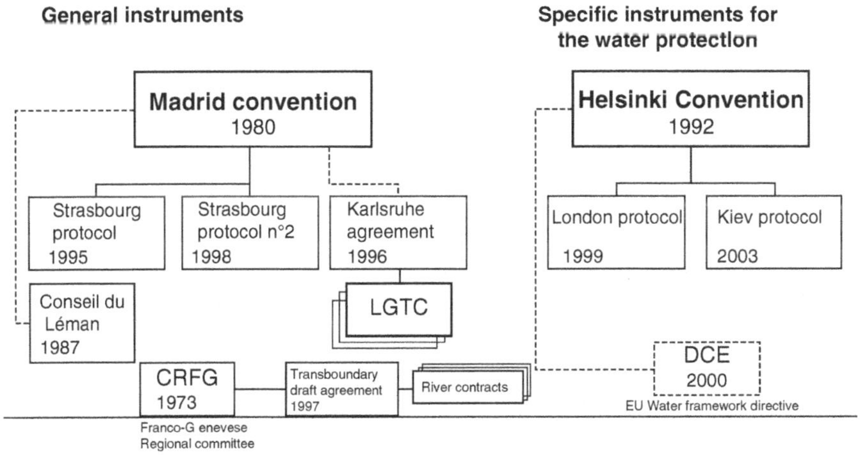

Fig. 7.7 Transboundary cooperation instruments

cross border cooperation instruments, for the simple reason that there were no such instruments in existence at the time. Several instruments of this type appeared as of 1980 and were ratified by both Switzerland and France (Fig. 7.7). There were three conventions on which a new agreement, governing the use of the Geneva aquifer, could be based, namely:

The European Outline Convention on Transfrontier Co-operation Between Territorial Communities or Authorities (the Madrid convention) of 21 May 1980 (entry into force 21 December 1981); the Madrid convention contains a standard inter-state agreement on transboundary cooperation among local authorities and a standard contract for service supply among "public law" cross border local communities.

The Convention on the Protection and Use of Transboundary Watercourses and International Lakes of 17 March 1992 (the Helsinki convention); the Helsinki convention lays down the obligations of States parties to cooperate without directly creating institutions or operational structures. The transboundary river contracts, to which Geneva associated itself, after signing the transboundary memorandum of understanding in 1997 to revitalize Franco-Genevese rivers, might have had its origins in this convention.

The Karlsruhe agreement on transboundary cooperation between local communities and local state entities of 23 January 1996; the Karlsruhe agreement, for its part, provides for the establishment of operational structures between local communities and/or local public authorities with legal personality.

These agreements do not have the same scope, as the Madrid Convention and its Strasbourg and Karlsruhe resulting protocol and agreement being regional do not have the same significance as the Helsinki UNECE International Convention.

Before choosing a transboundary cooperation instrument, it was necessary to decide on the objective to be achieved. If the aim was to replace the 1978 arrangement, a Local Grouping of Transboundary Cooperation (LGTC) (bringing together the local communities of Geneva, the State of Geneva and SIG) seemed to be the way forward. However, when it came to water management, the French entities and the State of Geneva would be the main interlocutors, with SIG's role restricted to artificial recharge exploitation. In fact, a simple solution, which was limited to groundwater use and recharge, was advanced in the form of a transboundary agreement among the local communities concerned. Based on the Karlsruhe agreement, particularly the possibility stipulated in its article 5 of delegating a mission to one of the local communities, an agreement was subsequently reached between the State of Geneva and the three French communities involved (the greater Annemasse region, the Community of Genevese communes and the commune of Viry). This option was considered as more appropriate than setting up an LGTC. Furthermore, in accordance with the Swiss federal constitution, cantons could sign treaties with foreign bodies within the areas of their competence. Cantons were thus free to act autonomously and themselves conclude international agreements of this type. Consequently, an agreement relating to the use, recharge and monitoring of the Franco-Swiss Genevese aquifer was signed between the communes of the greater Annemasse region, the Genevese communes and the commune of Viry on one side, and on the other, the State Council of the Republic and the canton of Geneva, on 18 December 2007. This new agreement replaced the 1978 arrangement and entered into force on January 1, 2008 for 30 years. The agreement or convention is a rare example of a transboundary aquifer management agreement between a Swiss canton and European Union communities. Hence, hereafter we give its main features:

The objective of the convention is to continue to ensure sustainability in the use of the aquifer's water resources for drinking water supply both in France and in Switzerland and to upgrade aquifer management considering recent legal developments.

The convention confirms the commission for the exploitation of the Geneva Aquifer, established in 1978, as an aquifer development commission, which encompasses two delegations—one for each country—of three members each, holding an ordinary session at least once a year, and whose functions are as follows:

- To define the annual groundwater utilization programme;
- To propose groundwater protection and conservation measures, as needed;
- To provide advice on the construction and modification of groundwater abstraction works;
- To control investment and exploitation costs; and
- To approve the foreseeable volumes of groundwater to be abstracted in each member state: these volumes will be considered as 'reserved.' France has decided not to abstract more than a specified annual volume.

The convention provides for the inventory and updating of groundwater recharge and utilization works. These works are to be equipped with devices to record abstractions and level variations, and these data are to be provided to the Commission. It further provides for the monitoring of water levels and water abstractions. Monitoring results are to be communicated to the Commission. It also calls for the regular analysis of water quality and the establishment of a warning system against accidental pollution. Finally, the convention contains provisions relating to the determination and allocation of the costs of aquifer exploitation and artificial recharge.

Meanwhile, the recharge plant, which, from the outset, had been placed under the responsibility of SIG for technical design and operation, was included in the transfer of assets between the State of Geneva and SIG. The asset transfer project consisted of transferring fixed assets belonging to the State of Geneva to public autonomous establishments. The transfer was both industrial and financial in nature with SIG allowed to assume complete control of the activity it had previously executed under State authority. SIG thus became the owner of the recharge plant as the State preferred to deal solely with sovereign tasks. As far as the recharge plant was concerned, in addition to the purely accounting and legal aspects, it was also necessary to provide guarantees for the agreements that had already been signed and those yet to be concluded between the State of Geneva and the French authorities (extension of the 1978 arrangement and the new agreement being drafted). Therefore, the agreement for the use and maintenance of the recharge facilities and related works between the State of Geneva and SIG needed to be amended. The amendment led the State of Geneva to concede the use of the artificial recharge plant to SIG. Lastly, whereas SIG used to pay annual rentals for the use of the Geneva State-financed plant, once the assets were transferred at their book value, annual rentals in SIG's accounts were replaced with financing costs and accounting depreciation. The law on asset transfer came into force on January 1, 2008.

In 30 years of operation, the Geneva artificial recharge system contributed over 250 million m^3 of water filtered from the Genevese water table. The choice made in the 1970s between a new plant for treating lake water and the creation of a recharge plant proved to be a judicious one, not only from a financial point of view, but also in terms of water resources management. Transboundary groundwater management, coupled with the technical success of the aquifer recharge system, have guaranteed safe drinking water for Geneva and the surrounding region by diversifying and optimising the quantitative and qualitative potential of existing water resources. Due to the radical change in demographics over the past years and the economic attractiveness of the Geneva region, there is an ever-greater trend towards establishing a cross border approach and a Franco-Valdo-Genevese territory regional project has been signed in December 2016, which will serve to create an environmental common strategic vision across the territory.

7.2.2 *The Guarani Aquifer System in South America*

The Guaraní Aquifer System or GAS (SAG is its acronym in Spanish) underlies Argentina, Brazil, Paraguay and Uruguay, with a total surface area of about 1.2 million km^2 (Fig. 7.8) (Villar et al. 2013). On 8 July 2004, the presidents of the four countries signed a Declaration through which they stated that, given the importance

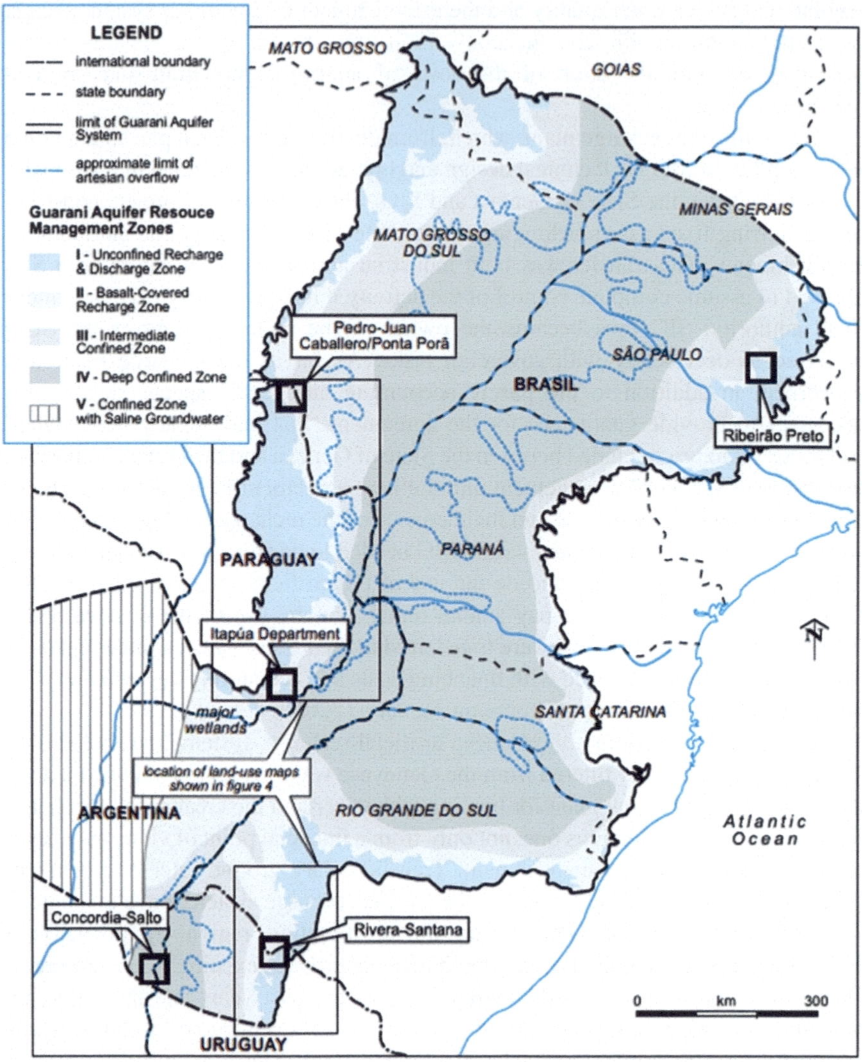

Fig. 7.8 Guarani aquifer system and management zones (Source: GWMATE 2009, p. 11)

of the System for the economy, all issues relating to it were to be dealt with at the level of MERCOSUR.[1] They created a High-Level Ad-Hoc Group and vested it with the task to formulate an agreement on the subject.

[1]Created in 1991, Mercosur is an accord between Argentina, Brazil, Paraguay and Uruguay calling for "the free movement of goods, services, and factors of production between countries"

In parallel, a GEF-sponsored project was launched to assist the system states in the pursuance of the long-term objective of creating a common legal, institutional and technical framework for the system's management and preservation. Given that the legal and institutional aspects were dealt with in parallel by the Ad-Hoc Group, the project focused on technical issues and on localized problems through pilot projects.

An international agreement was developed by the Ad-Hoc Working Group, and signed in August 2010 by Argentina, Bolivia, Brazil and Paraguay, but not yet in force. This Agreement follows the main guidelines of the United Nations Law of Transboundary Aquifers, especially in relation to the following principles: sovereignty, the equitable and reasonable use of water resources, the obligation not to cause harm, cooperation, and the exchange of data and information (Vilar and Ribeiro 2013). Its main features are the following:

- The parties have sovereign territorial control over their respective portions of the GAS;
- The parties undertake to promote the management, monitoring and sustainable development of GAS water resources within their territories;
- The use of the resources of the GAS shall be based on criteria of rational and sustainable use and on the duty of the aquifer countries not to cause appreciable harm to other aquifer countries or to the environment;
- The parties shall promote the conservation and protection of the GAS;
- Activities with possible adverse effects shall be planned in the respect of prevailing international law. In particular, the parties shall notify each other any planned activity or work with possible transboundary effects, and in case of objections, the parties shall consult each other with a view to reaching agreement on an equitable solution;
- The parties have the duty to eliminate or reduce the damage caused, if any;
- The parties shall exchange data and information on the GAS, as well as on activities, studies and works relevant to the sustainable development of the system;
- The parties shall cooperate in the identification of critical areas (hot spots);
- Disputes shall be settled through direct negotiations; failing these, resort shall be had to the commission for an arbitration procedure.

Finally, Article 15 states that a dedicated multilateral Commission will oversee and coordinate the cooperation process and facilitate the implementation of the Agreement (Vilar and Ribeiro 2013). Unfortunately, the countries have yet to establish it and determine its statutes, competences, members and budget. Desirable objectives for the Commission would be: leadership in disseminating and producing knowledge about the aquifer; harmonization of legal instruments such as wellhead protection areas and groundwater permits; establishment of methodological guidelines for a groundwater database; and coordination of a common groundwater informational system. As the agreement did not mention the recharge areas of the aquifer with higher natural vulnerability that are more likely to create conflicts, the Commission could take the lead in designing a common strategy to manage these areas, especially the ones within or very near the frontier zone. In case of conflicts over the use of the Guarani Aquifer, the Commission would be in a position to present recommendations.

Article 17 affirms: "If through direct negotiations an agreement is not reached within a reasonable period, or if the dispute is only partially resolved, the Parties in the controversy shall, through mutual agreement, solicit the Commission related in Article 15 to, upon a presentation of the respective positions, evaluate the situation and, if appropriate, formulate recommendations."

According to this article, however, the Commission will still have a restricted role because its participation has to be evoked by the Parties through mutual consent and its intervention has no binding consequences. If the countries can't reach an agreement after this procedure, Article 19 mentions the possibility of an arbitration procedure which would be defined by a future protocol.

It has been debated whether an autonomous institutional mechanism for cooperation in the management of the GAS should be created, or whether aquifer management should take place within the existing framework of the Intergovernmental Commission for Coordination (ICC). Since the GAS, although somehow related to the Plata basin, is largely independent of surface water flows, it would perhaps be more logical to have a separate cooperation mechanism, though coordinating with the ICC.

It is also interesting to underline that some researchers have suggested that the GAS is as much a political construction as a hydrogeological reality of interconnected aquifers and that it answers some political objectives. This idea has been defended by Walter (2015): "Ideas played a key role in the development of the GAS policy. Indeed, the notion of a transboundary groundwater body in the Southern Cone only emerges in the final decades of the twentieth century. Before then, although the existence of aquifers in the region was well known, scientists had failed to identify the transnational linkages between groundwater bodies. The aquifers that make up the GAS were studied as separate entities under different names... the name "Guarani" was chosen for its sociocultural implications instead of a more conventional nomenclature... The idea of the GAS was scientifically controversial because the geological formations that make the GAS are extremely dissimilar across the region, but it was also politically compelling. The existence of a unique system of connected aquifers implied that groundwater exploitation decisions were interdependent. It meant that groundwater exploitation practices were having an impact beyond jurisdictional borders, and that some degree of international coordination would be necessary to adequately exploit the resources in the medium to long term".

Walter makes an interesting comment, which opens some new perspectives for hydro-governance: "The processes of social construction that lead to the recognition of groundwater problems are too often ignored by the literature dedicated to water governance. Problems tend to be taken for granted—seen as existing *a priori* of the policy-making process—or, alternatively, framed as purely instrumental to the strategic preferences of political actors".

Finally, although the agreement did generate rather great expectations among the countries which signed it, a lack of practical implementation and the absence of genuine cooperation among these countries has handicapped a promising process. Yet, to end this presentation on a positive information, a Regional Centre for Groundwater Management for Latin America and the Caribbean (CeReGAS) has been created in Uruguay as a partnership between that country and UNESCO/IHP, to contribute to the improvement of knowledge about groundwater in that part of the world and stimulate the concerned countries to cooperate on a common water resource.

To conclude, from a governance perspective, the two ingredients of a sound hydro-governance are at least present: an International Agreement and an International River Basin Organization (IRBO), the multilateral Commission. But, while the Agreement is quite ambitious, its power remains limited: it has not yet been ratified by all partners, Brazil for instance, the Commission has not been clearly established and a dispute resolution mechanism has to be set. There is still quite a distance from theory to practice (Fig. 7.9).

7.2.3 The San Pedro Aquifer Between the US and Mexico

There is a general agreement between the US and Mexico concerning surface waters and, more specifically, the rivers Colorado, Tijuana and Rio Grande, established in 1889 and adjusted by treaty in 1944 with the creation of the International Boundary and Water Commission (IBWC). But it should be noticed that the case of groundwater has not been accounted for politically and legally speaking, with a few exceptions. But, due to the evolution of climatic circumstances, groundwater has become a significant water resource for the region on both sides of the political border and the scientists do realize that significance, as illustrated by the Binational Study of the Transboundary San Pedro Aquifer, which mainly concerns groundwater quality and emphasizes the science and data harmonization drivers. Its final report has been published in 2016 (Callegary et al. 2016) and we quote it hereafter to illustrate the role of scientific cooperation in transboundary groundwater governance.

"To facilitate transboundary scientific cooperation and coordination, the binational Transboundary Aquifer Assessment Program (TAAP) was officially launched on August 19, 2009, with the signing by the Mexican and U.S. Principal Engineers of the IBWC (International Boundary and Water Commission) of the "*Joint Report of the Principal Engineers Regarding the Joint Cooperative Process United States-Mexico for the Transboundary Aquifer Assessment Program*". This document serves as the framework for U.S.-Mexico coordination and dialogue to implement transboundary aquifer studies. The document clarifies several details about the program such as background, objectives, roles and responsibilities, funding, relevance of the international water treaties, and the use of information collected or compiled as part of the program. Sharing data and information in both countries creates the basis for cooperation and a common understanding in support of

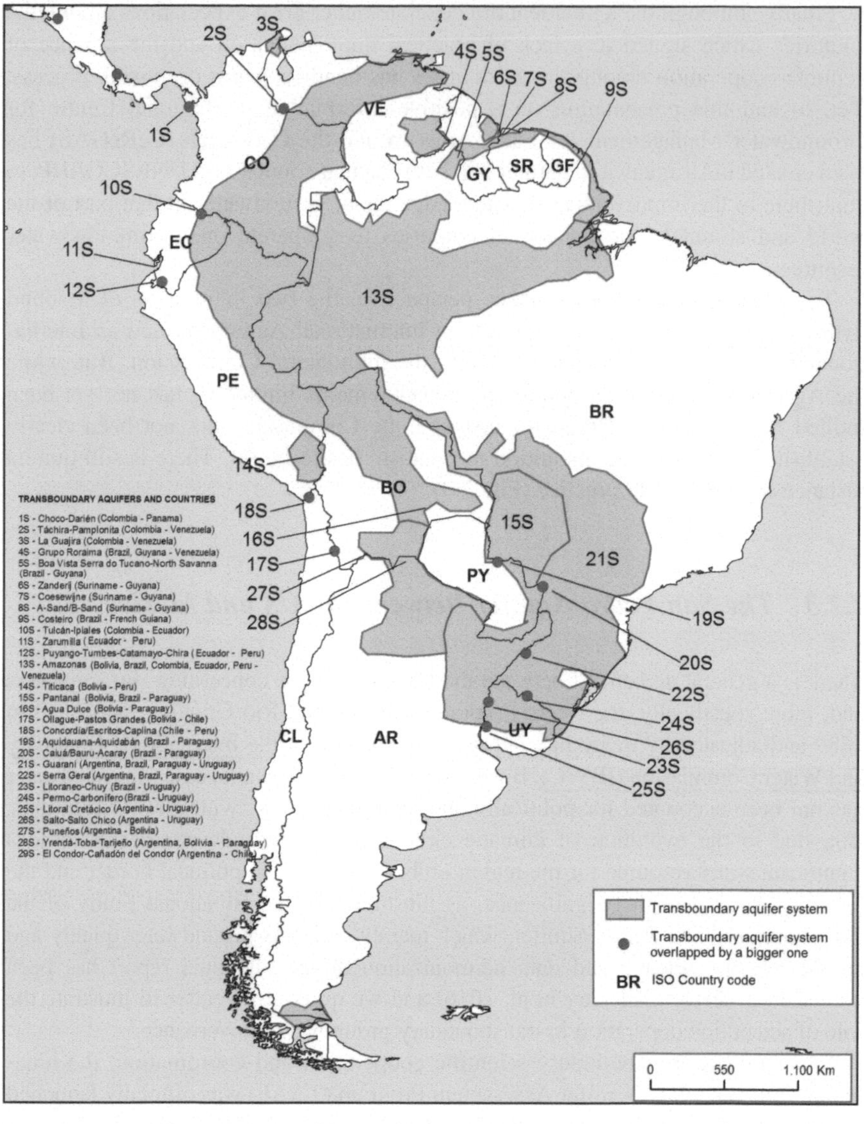

Fig. 7.9 Transboundary aquifers in South America (Elaborated by Villar 2015)

sustainable use and protection of water resources and achieving sustainable economic development. In addition, the intellectual and interpersonal interaction improves the relationships needed to coexist in harmony at a local and international level.

The technical team for this study of the Arizona-Sonora aquifers included personnel from the two federal governments (CONAGUA and USGS) and two state universities (University of Sonora-UNISON and University of Arizona-UA) working together under the auspices of the IBWC. The participants in this Arizona-Sonora effort also included technical expertise from the faculty and staff of the Geology Department at the UNISON and the UA Water Resources Research Center, enabling a significant advance in the binational understanding of the basin. The participation of the two universities incorporated an academic aspect that reflected a culture of transparency and openness common to academia around the world. This academic collaboration brought flexibility to the information exchange and the development of a framework of mutual understanding that benefits both federal governments.

Considering the basin characteristics described in this study, binational information was available to estimate the parameters of the regional climate and vegetation system, but there were no previously existing binational integrated maps with information for soils, geology, geophysics, piezometry, hydraulic parameters, or water quality. Differences on both sides of the border in hydrography, hypsometry, and terrain slopes were, at least in part, caused by differences in the resolution of digital elevation models. One of the important results of the study was to create binational maps and data sets for many of the data types needed for a regional hydrological analysis. The surface geology and geophysical information—electromagnetic, gravimetric, magnetic, and seismic—have been used to develop the conceptual model of binational hydrogeology. The available data for climate, hydraulic parameters, piezometry, and hydrogeology may be used to update a groundwater flow model. This process has also served to identify gaps in data and the importance of monitoring on a binational level. Notably, the information on soil types could not be integrated on a transboundary basis because each country uses a distinctly different classification system. This joint, uniform integration would require a complex, long-term project requiring bilaterally negotiated methods and the participation of a binational field team specializing in soil mapping. With this exception, the integration of transboundary hydrogeological data was possible although sometimes requiring joint interpretation and dialogue among scientists from both countries".

In Sect. 5.2.2, we already briefly presented this case as an illustration of **the paradox of transboundary hydro-governance without a formal treaty.**

A summary of the results of the Final Report has been given by Megdal (2017) during the TAAP forum in Sierra Vista AZ in June 2017, as the Cooperative Framework for the TAAP, listed in the chapter references. Some points of general interest for the hydro-governance of transboundary aquifers are the following (Megdal 2017):

"The Cooperative Framework acknowledges the need to develop a team of binational experts to assess aquifers, exchange data, and if necessary, develop new datasets. Each country will be responsible for any costs on projects conducted in its territory, in addition to selecting the participants and consultants to carry out the studies in that country. Each country may contribute to costs for work done in the other country, and the IBWC will coordinate any flow of funds across the border."

"The six principles of agreement, which appear toward the end of the three-page document, make it clear that each country is free to undertake its own studies when such are limited to one side of the border (Fig. 7.10). They are as follows:

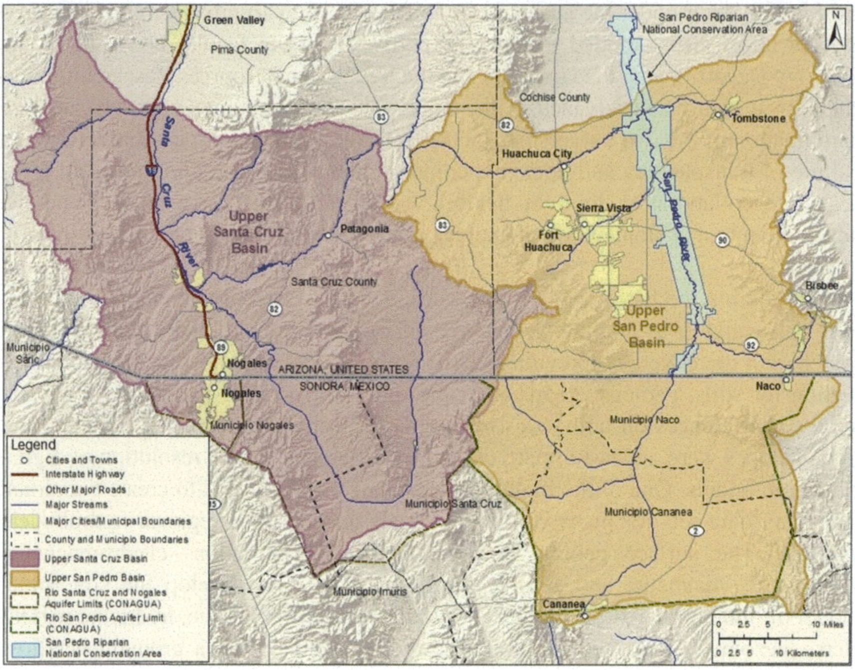

Fig. 7.10 San Pedro River and Santa Cruz River binational basins (Binational Study of the San Pedro Aquifer-Final Report 2016)

1. Activities described under this agreement should be beneficial to both countries.
2. Aquifers to be jointly studied, as well as the scope of the studies or activities to be done on each aquifer, should be agreed upon within the framework of the IBWC.
3. The activities should respect the legal framework and jurisdictional requirements of each country.
4. No provisions set forth in this agreement will limit what either country can do independently in its own territory.
5. Nothing in this agreement may contravene what has been stipulated in the Boundary and Water Treaties between the two countries.
6. The information generated from these projects is solely for the purpose of expanding knowledge of the aquifers and should not be used by one country to require that the other country modify its water management and use".

7.2.4 The Kilimanjaro Aquifer Between Kenya and Tanzania

The Kilimanjaro Aquifer is characterized as a volcanic aquifer system, whose catchment area is Mount Kilimanjaro, with a recharge by rainfall and melting ice. The recharge is receding due to a decrease of the areal extent of Kilimanjaro's ice cap of about 80% since 1912. Groundwater flows towards both the Kenya and Tanzania sides of the mountain, but in variable quantity (Mwango n.d.) and quality. Two lakes form part of the Kilimanjaro aquifer area: Lakes Jipe and Chala. Lake Jipe is a small and shallow transboundary lake, whose area has declined over the years; this has been attributed to catchment degradation, leading to erosion and siltation (Mwango n.d.). The Lumi River comprises the most significant drainage system in the Lake Jipe watershed. The headwaters of the river start from the southeastern slopes of Kilimanjaro and form small tributaries. These tributaries remain dry most of the year and the perennial flow of the Lumi River is maintained by a system of high yielding springs (Mwango n.d.). A series of these springs join Lumi River until it empties into the Lake Jipe (ESF 2005; Mwango n.d.). Lake Chala, which is to the North of Jipe, is a small freshwater crater lake 85–90 m deep. Based on deuterium and oxygen-18 components in Lake Challa's waters, it has been assumed that the springs in the area around the lake do not get a major part of their waters from the lake, and two springs show a high probable partial recharge by the lake water (Payne 1970).

Like in all other parts of the region, the area is experiencing high population growth rates and an increasing demand for water and other land-based resources. Water is used for irrigation, domestic purposes, and livestock and wildlife watering. It is important to have ecosystems and biodiversity conservation that supports tourism and recreation, as well as fisheries, and transport (as a cross-border crossing point). Sand harvesting takes place along river ways, especially River Lumi. Forestry, and HEP generation on River Ruvu in Tanzania are also important. One of the remarkable contemporary features of the landscape on the slopes of Kilimanjaro is the sharp land-use cover boundary that cuts across the gradient, coinciding with the political boundary of Kenya and Tanzania (Campbell et al. 2004). Both ecological and socio-economic conditions define this context, the socio-economic one being influenced by patterns of land use (farming, herding, wildlife) interactions and exchanges between them, and particularly their relationships with broader regional, national, and international processes such as trade and migration, and economic, social and environmental policies (Campbell et al. 2004).

There are about 150 groundwater sources including boreholes and shallow wells on the Kenya side (ESF 2005). These are mainly used for domestic, small-scale irrigation and livestock watering. The boreholes in the Taveta division have been drilled into the volcanic aquifer system, which is characterized by high ground water potential due to high permeability of the pyroclastic formations and storage at the volcanic/basement interface (ESF 2005). Groundwater tables in some areas of Taveta are to be found at relatively shallow depths between 0.5–2 m below surface. This is also the range at which most of the shallow wells that form the bulk of ground water resources exist. From about 20 m below surface, the calcareous formation becomes impervious preventing percolation of water into deeper aquifers.

Problems affecting surface and groundwater supplies include: lack of an ecosystem-integrated management plan; environmental degradation; cross-border conflicts due to declining resources, water diversion, etc.; accelerated run-off; increasing siltation of rivers; deteriorating water quality due to chemical pollution from irrigated lands; a collapsing fishery industry; and altered ecology.

The governments on both sides of the border have designed and implemented a variety of policies and related legislation that have influenced patterns on land use. While the administrative structure in Kenya has been relatively consistent, that in Tanzania experienced significant changes particularly after independence (Campbell et al. 2004).

In Tanzania for example, the definition of land property rights changed from customary to government owned and now to freehold (Campbell et al. 2004). The Kenya National Water Policy is silent on Transboundary Water Management (Mwango 2009). Kenya's Transboundary Water Resources Management and Development is encapsulated in Sessional Paper No 1 of 1999 while its national law on water resources management is the Water Act, 2002, which has put in place a comprehensive framework for the management and sustainable utilization of water resources (Mwango n.d.), which deals with issues relating to water resources management, water supply and sewerage development, institutional frameworks, and financing of the sector (Mwango n.d.). The Tanzania National Water Policy recognizes that the country is riparian to transboundary water resources with neighboring countries. It notes that these water bodies, which include Lakes Victoria, Tanganyika, Nyasa, Chala and Jipe, as well as Kagera, Mara, Umba and Songwe Rivers, exhibit unique characteristics and a complex range of water management and development issues and challenges. However, it is also silent on groundwater issues, as it is in Kenya.

Given the economic and environmental situation, sustainable management of the transboundary area between Amboseli Biosphere Reserve and Kilimanjaro World Heritage has become a major challenge affected by climate change impact (UNESCO 2009). Unlike transboundary surface water and river basins, transboundary aquifers are not properly documented and presently there are no agreements or Memorandums of Understanding on Transboundary Aquifers, which are managed by each respective country. There is no joint data base and scientific correlation of the hydrogeology of shared aquifers is deficient, also including scarcity of regional information and gaps in resources potential (quantity and significance to socio-economic development and ecosystems) (Mwango n.d.). Issues related to shared aquifers and sustainable use remain blurred because of poorly developed institutions, a lack of capacity and awareness (Mwango n.d.).

Groundwater has still a rather low profile, like in many other regions of the world, partly because it is a largely invisible resource, and yet it is heavily exploited particularly during drought periods. To better manage and use the transboundary groundwater resources, stakeholders and the public need to become aware of the need for and importance of groundwater in relation to their welfare and livelihoods, as well as its importance to the environment (ecosystems and services). More generally, there is a need to integrate aspects of national water policies that impact such resources, as well as harmonize rules and regulations for their management in a consistent and equitable manner across borders. The profile and role of groundwater in the regional sustainable development agenda should be greatly upgraded through effective regional institutions, even joint institutions, and agreements for the joint management of transboundary waters, surface and groundwater, active and meaningful stakeholder involvement across the borders, associated with education and awareness programs enhancing human resources capacities across borders, completed by a shared transboundary water database.

It should be stressed that transboundary preoccupations and the political and legal instruments to account for them have been partly approached in the case of surface waters by Kenya and Tanzania, whose governments agreed upon a joint transboundary management programme of the ecosystems (Pangani Basin) comprised of Lakes Chala and Jipe and the Umba River, signing a memorandum of understanding (MoU) during the session of the coordination committee of the eleventh sectoral council of ministers for Lake Victoria Basin that took place at Kisumu, Kenya, in February 2013, (Muwanga 2013), whose objectives are to establish a Joint Cooperative Framework for sustainable development and management of the ecosystems of the Basins and to set up an institutional arrangement for the management of ongoing projects, programmes and initiatives in the Basins (Pangani 2015).

Joint transboundary water management is led by the Pangani Basin Water Board, whose responsibilities include: Water Resources monitoring and assessment—Water Allocation (issuing and management of water permits)—Strengthen community participation in Water Resources Management—Coordinate water resources management and development planning—Water quality monitoring and pollution control—Water use conflict management—Water sources protection and conservation. It should be stressed that, from the political and legal points of view, in Tanzania, the National Water Policy (2002) makes explicit provisions for the management of transboundary water resources, and in Kenya, the Water Act (2002) has created the Water Resources Management Authority and provided a role for user groups, organized as Water Resources Users Associations (WRUAs), in the management of water resources (Fig. 7.11).

Otherwise, a Dutch-sponsored and funded program Kenya-Water 2015, organized by the Dutch Embassy in Nairobi, has worked on transboundary and collective river basin management, drawing the following conclusions, which illustrate the state of transboundary water policy and management in Kenya and neighbouring countries: "Kenya's shared water resources accounts for over 50% of the country's renewable surface water resources. Kenya shares surface and groundwater with all its neighbours, this includes 10 international trans-boundary river basins and lake basins. The government has over the past years been developing its Transboundary Water Policy. By the end of 2014 the policy was approved by the Attorney General (AG) and the Ministry of Foreign Affairs and International Trade (MFA). Kenya is presently finalising the signing of bilateral MoUs with the neighbouring countries Ethiopia, Tanzania (see above), and Uganda. Kenya has been taking part in negotiations around the Nile basin, and is in the process of ratifying the Nile Corporative Framework Agreement. Kenya and Ethiopia in collaboration with UNEP have initiated a regional project on the sustainable management of Lake Turkana and its River Basins. The project document has been signed by the two countries (and by UNEP in 2015)...."

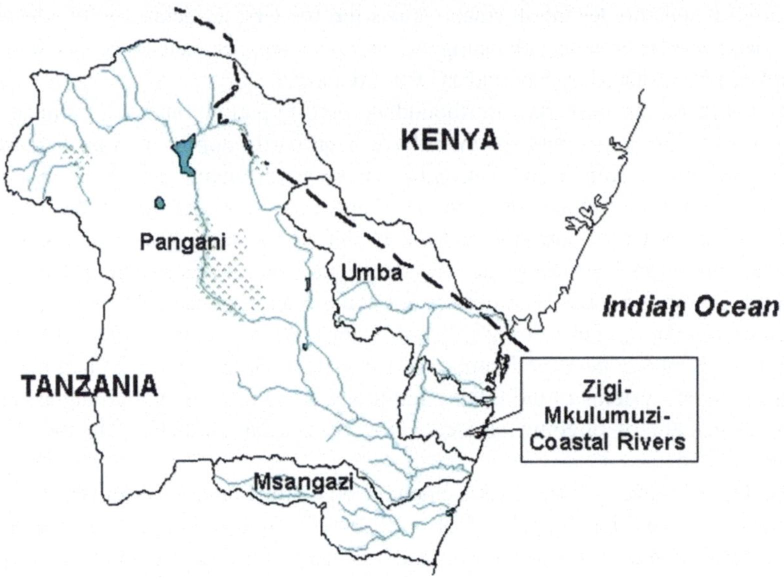

Fig. 7.11 The Pangani basin

To conclude, the path taken by Kenya and its neighbours to manage their common waters and illustrated by the Dutch study involves, on the one hand, the designing and signing of a common transboundary policy document and, on the other hand on the practical side, the creation of an IRBO, International River Basin organization, as illustrated by the case of the Pangani Basin and its Water Board. It seems likely, but has still to be done, that this path could be followed where groundwater is concerned (Olago 2016).

7.2.5 The Al-Disi Aquifer Between Jordan and Saudi Arabia

The Al-Disi aquifer, located south of Jordan, expands over 4000 km^2 in Jordan, estimated at a volume of 10 km^3, and about 65,000 km^2 in Saudi Arabia, estimated at a volume of 65 km^3, where it is known as the Al-Sag aquifer (Fig. 7.12). It is vital as its water is fossil, because there is no direct replenishment of the aquifer via rain due to the lack of rain in its area and the high evaporation rate. The aquifer's recharge process is only partial and is restricted to the Um Sahem and Ras Al Naqab areas in Jordan. The aquifer's fossil water dates to 10,000–30,000 years back and has been stored when the area witnessed plenty of rainfall back then. The issue between the two countries has been the over pumping of the aquifer by Saudi Arabia for its wheat production, yielding a reversal of groundwater flow from Jordan to Saudi Arabia, while Jordan was also increasing its withdrawals because of the development of its agriculture along its southern border. Besides, Jordan had to face water scarcity in

Fig. 7.12 The Al-Disi aquifer

other regions and designed, built and started operating the Disi Water Conveyance Project, a pipeline from Disi to Amman, and extracting a rather considerable extra amount of water from the Al Disi aquifer (Ellingson et al. 2011).

To try and solve these issues, Jordan and Saudi Arabia decided two actions: the adoption and signature of an **Agreement**; and the creation of a **Joint Technical Committee**. Hence, in 2015, Jordan and Saudi Arabia did sign an Agreement governing the "sound management and protection of the shared Al-Disi aquifer", which supplies the capital of Jordan, Amman, and other governorates with a quarter of their water needs, according to government officials (Disi_Aquifer_Agreement-English 2015). The Agreement is concise with four main articles:

- Article One-contains terms and definitions;
- Article Two-describes the main norms for managing the aquifer
- Article Three-discusses the creation and responsibilities of a Joint Saudi/Jordanian Technical Committee
- Article Four-contains administrative provisions related to the implementation of the Agreement.

Besides, the agreement identified a 10 km buffer zone on both sides of the Jordanian-Saudi borderline, a "protected (or prohibited) area" where drilling wells for producing water is prohibited, meaning that all activities depending on the extraction of groundwater should be eliminated or forbidden, while wells drilled beyond the buffer zone (or Management Zone) must adhere to shared technical standards (Areas defined of a map attached to the agreement, the Map for the Agreement Concerning the Management and Utilization of the Ground waters in the Al-Sag/Al-Disi). In addition, both countries agreed to take all necessary measures that prevent the contamination of the aquifer to ensure the safety and sustainability of the shared Jordanian Al-Disi aquifer and the Saudi Al-Sag aquifer.

A joint Jordanian-Saudi **technical committee** was formed to oversee the implementation of the agreement, with the following tasks: to supervise the implementation of the terms of the Agreement; to supervise and observe extracted groundwater quantity and quality; to collect and exchange information, statements and studies and their analysis, and submit the results to the competent authorities in Saudi Arabia and Jordan. Members of the committee will meet every 6 months or when the need arises.

The agreement limits the use of groundwater to municipal use, and its use for irrigation should not be authorized any more, which seems quite difficult to implement on the Saudi side. To try and solve this issue, an interpretation could be that only new uses of the aquifer should be limited to municipal purposes (Eckstein 2015).

Al-Disi aquifer water is a main water source for drinking and industrial purposes in Amman, Zarqa and Aqaba, and work is under way to transfer Disi water to the northern governorates. Hence, the agreement also protects water projects which benefit from the Al-Disi aquifer, including the Disi Water Conveyance Project, already mentioned, which conveys 100 million m^3 annually from the Disi Aquifer to Amman and started its operation in 2013.

As a conclusive comment, the experience of Al Disi shows that, although the development and management of shared water resources between Jordan and its neighbouring countries is a challenge for the water professionals and diplomats, there could be hope that these shared waters will contribute to the political stabilization of the region.

References

Binational study of the transboundary San Pedro aquifer-Final Report (2016) International boundary and water commission. Universidad de Sonora, CONAGUA, University of Arizona, USGS

Burchi S, Spreij M (2003) Institutions for international freshwater management. UNESCO/IHP/WWAP IHP-VI, Technical Documents in Hydrology/PCCP series n°3. http://unesdoc.unesco.org/images/0013/001324/132478e.pdf

Callegary JB, Minjárez Sosa I, Tapia Villaseñor EM, dos Santos P, Monreal Saavedra R, Grijalva Noriega FJ, Huth AK, Gray F, Scott CA, Oroz Ramos LA, Megdal SB, Rangel Medina M, Leenhouts JM (2016) Binational study of the transboundary San Pedro aquifer-final report. International Boundary and Water Commission, Universidad de Sonora, CONAGUA, University of Arizona, USGS

Campbell DJ, Misana SB, Olson JM (2004) Comparing the Kenyan and Tanzanian Slopes of mount Kilimanjaro: why are the neighboring land uses so different? LUCID project working paper 144. LUCID Project, International Livestock Research Institute, Nairobi, Kenya de los Cobos G (2002) The aquifer recharge system of Geneva (Switzerland): a 20 year successful experience. In: Dillon PJ (ed) Management of aquifer recharge for sustainability. AA Belkama Publishers, Lisse, pp 49–52

de los Cobos G (2007) Impacts of a long-term shutting down on the aquifer recharge management: case of the aquifer recharge of Geneva, Switzerland. In: Fox P (ed) Management of aquifer recharge for sustainability. Acacia Publishing Incorporated, Phoenix, pp 296–306

de los Cobos G (2016) The transboundary aquifer of the Geneva region (Switzerland and France): successfully managed for 30 years by the state of Geneva and French border communities. In: Fried J, Ganoulis J (eds) Transboundary groundwater resources: sustainable management and conflict resolution. Lambert Academic Publishing, Saarbrücken, pp 183–193

Disi_Aquifer_Agreement-English (2015).pdf https://www.internationalwaterlaw.org/documents/regionaldocs/Disi_Aquifer_Agreement-English2015.pdf

East African Community Victoria Basin Commission Secretariat: Draft Report on Consultancy Services to: Harmonization of Regional Policies, Rules and Regulations, Governing Utilization of Water Resources in the Lake Victoria Basin-WEMA Consult Ltd. www.wemaconsult.com

Eckstein G (2015) The newest transboundary aquifer agreement: Jordan and Saudi Arabia cooperate over the Al-Sag/Al-Disi Aquifer. https://www.internationalwaterlaw.org/blog/2015/08/31/the-newest-transboundary-aquifer-agreement-jordan-and-saudi-arabia

Ellingson J, Abadesco E (2011) The Disi-Amman water conveyance project. http://courses.washington.edu/cejordan/Disipresentation.pdf

ESF (2005) Baseline survey report for Lake Jipe: emergency community-based recharge interventions for the restoration of Lake Jipe. Supported by EU – BCP/CDTF Programme. Environmentalistes Sans Frontieres Consultants

GW MATE (Groundwater Management Advisory Team) (2009) The Guarani aquifer initiative – towards realistic groundwater management in a transboundary context. The World Bank, GW MATE and Water Partnership Program. Case number 9. http://siteresources.worldbank.org/INTWAT/Resources/GWMATE_English_CP_09.pdf. Available 19 Dec 2017

International Commission for the Protection of the Danube River. https://www.icpdr.org/. Accessed 19 Dec 2017

International Joint Commission (IJC) (2017a) 2015–2020 Strategic plan. http://www.ijc.org/en_/Strategic_Plan Accessed 19 Dec 2017

International Joint Commission (IJC) (2017b) More than a century of cooperation protecting shared waters. http://www.ijc.org/en_/Role_of_the_Commission. Accessed 19 Dec 2017

International Waters Governance (2011) http://www.internationalwatersgovernance.com/danube-river-basin.html. Accessed 19 Dec 2017

International Waters Governance (2018) R Paisley. http://www.internationalwatersgovernance.com/senegal-river-basin.html

International Waters Governance Guarani Aquifer System. http://www.internationalwatersgovernance.com/guaraniacute-aquifer-system.html. Accessed 19 Dec 2017

International Waters Governance, the Rhine. http://www.internationalwatersgovernance.com/the-rhine.html

Jensen KM, Lange RB (2013) Transboundary water governance in a shifting development context new development finance development spaces and commitment to cooperation: a comparative study of the Mekong and the Zambezi River Basins DIIS Report 2013:20

Kenya-Water (2015) Rijksoverheid. https://www.rijksoverheid.nl/binaries/rijksoverheid/documenten/rapporten/2016/09/08/kenya-water-2015/kenya-water-2015.pdf

Kinna R, Rieu-Clarke A (2017) The governance regime of the Mekong River basin: can the global water convention strengthen the 1995 Mekong agreement? Int Water Law 2(1):1–84

Megdal SB (2017) The cooperative framework for the transboundary aquifer assessment program: a model for collaborative transborder studies. Public Policy Review, TAAP forum in Sierra Vista, AZ, June 20, 2017, Arizona Water Resource. http://wrrc.arizona.edu/TAAP. Accessed 19 Dec 2017

Mekong River Commission. http://www.mrcmekong.org/. Accessed 19 Dec 2017

Mumme SP (2001) The US-Mexico international boundary and water commission in the sustainable development era IBRU boundary and security bulletin. http://citeseerx.ist.psu.edu/viewdoc/download;jsessionid=7DBC88A5C55FD8CC10E953DCB18C98DB?doi=10.1.1.525.6517&rep=rep1&type=pdf

Muwanga D (2013) East Africa: Tanzania, Kenya in joint effort to manage ecosystem. East African Business Week (Kampala) All Africa.com [26 Feb 2013]

Mwango FK (2009) Shared water resources: Kenyan case. Paper presented at World Water Week in Stockholm, Sweden, August 16–22, 2009. Ministry of Water and Irrigation, Kenya

Mwango FK (n.d.) Present state of Kenya transboundary aquifer resources (Legislation and database management). Ministry of Water and Irrigation, Kenya

Olago DA (2016) The transboundary aquifers of Kenya: their characteristics, status, management, and legislative, socio-economic and political contexts for their wise and sustainable use-case study of the Kilimanjaro aquifer. In: Fried J, Ganoulis J (eds) Transboundary groundwater resources: sustainable management and conflict resolution. Lambert Academic Publishing, Saarbrücken, pp 201–212

Pangani Basin Water Board (2015) Water resources management in Pangani basin. Training on International Water Law in Africa

Payne BR (1970) Water balance of Lake Challa and its relation to groundwater from tritium and stable isotope data. J Hydrol 11:47–58

Regional Stakeholder Forum on the Mekong Adaptation Strategy and Action Plan, Vientiane, Lao PDR, 2017. http://www.mrcmekong.org/news-and-events/events/regional-stakeholder-forum-on-mekong-adaptation-strategy-and-action-plan-26th-june-2017-vientiane-lao-pdr; http://www.mrcmekong.org/about-mrc/core-river-basin-management-functions/. Accessed 19 Dec 2017

River Basins Commissions and other Institutions for Transboundary Water Cooperation-Capacity for Water Cooperation in Eastern Europe, Caucasus and Central Asia series (2009) ECE/MP.WAT/32. http://www.unece.org/?id=11628

Rivera A (2015) Transboundary aquifers along the Canada-US border: science, policy and social issues. J Hydrol Reg Stud 4:623–643

Sène AM, Bonin S, Soubeyran O (2007) Watershed regulation and local action: analysis of the Senegal River watershed management by a regional organisation and public participation. Institut de Géographie Alpine, Laboratoire PACTE. http://www.hydrol-earth-syst-sci-discuss.net/4/1917/2007/hessd-4-1917-2007.pdf. Accessed 19 Dec 2017

The International Boundary and Water Commission 2011/2012 its mission, organization and procedures for solution of boundary and water problems binational border water resource summit. https://www.ibwc.gov/About_Us/About_Us.html. Accessed 19 Dec 2017

The International Boundary and Water Commission. http://www.ibwc.gov/about_us/About_Us.html. Accessed 19 Dec 2017

The Mekong River Basin Strategic Plan 2016–2020 Mekong river commission. http://www.mrcmekong.org/assets/Publications/strategies-workprog/MRC-Stratigic-Plan-2016-2020.pdf. Accessed 19 Dec 2017

Transboundary Cooperation Fact Sheets (2012) Part of "Comparative Study of Pressures and Measures in the Major River Basin Management Plans" Task 1 – Governance final report 28 Nov 2012. http://ec.europa.eu/environment/archives/water/implrep2007/pdf/Governance-Transboundary%20Fact%20Sheets.pdf. Accessed 19 Dec 2017

UNESCO (2009) Biennial report on UNESCO science activities by the field offices in Africa (2008–2009). UNESCO Nairobi Cluster Office and Regional Bureau for Science and Technology in Africa

Utton AE (1978) International groundwater management: the case of the U.S.-Mexican frontier. Nebr Law Rev 57(3):633

Villar PC, Ribeiro WC (2013) The agreement on the Guarani Aquifer: cooperation without conflict. Global Water Forum

Villar PC (2015) Aquíferos Transfronteiriços: Governança das águas e o Aquífero Guarani. Juruá, Curitiba

Walter M (2015) The invention of the Guarani aquifer system. Water and environment, ReVista, Harvard Review of Latin America. http://revista.drclas.harvard.edu/book/invention-guarani-aquifer-system. Accessed 19 Dec 2017

Chapter 8
Towards A "Good" Transboundary Hydro-Governance

Anyone who can solve the problems of water will be worthy of two Nobel prizes: one for peace and one for science.
John F. Kennedy, Former US President

Abstract Having introduced the why, what, how of Transboundary Hydro-Governance, this chapter proposes a detailed organized framework of "good" hydro-governance, i.e. a practical model based on seven drivers. It examines how they could be built, how they could be coordinated and how they could operate, from Science and Technique to Policy and Law to Socio-Economics to Diplomacy and Cooperation, taking Scaling into consideration (local, regional, national, international), introducing River Basin Organization and insisting on the role of Capacity-Building.

Following this chapter, we conclude the whole book by stressing that the framework we propose, derived from our personal experience, is practical and clearly is a progress with respect to the usual instruments of hydrogovernance. But we ask for caution as the benefits from international cooperation around transboundary watercourses have in most cases not materialized and even less for transboundary aquifers, and much work has still to be done.

Keywords Integration · Science · Economy · Social · Diplomacy · Drivers · Joint management · Capacity-building · Scaling

8.1 Transboundary Hydro-Governance: Why? What? How?

Why: the major feature of a transboundary water situation is that, on either side of the boundary, a given water resource should answer different social and economic development priorities that can be conflicting due to historic, political, geographical and environmental conditions and heritage and complicated by language, culture and education differences. This feature and the rather high number of transboundary

J. Ganoulis, J. Fried, *Transboundary Hydro-Governance*,
https://doi.org/10.1007/978-3-319-78625-4_8

water resources worldwide, either surface water or groundwater, justify that a specific model of hydro-governance be developed for transboundary situations, offering a general framework for risk and uncertainty analysis in conflict resolution, where risks and benefits could be shared between riparian countries and "win-win" solutions to transboundary disputes can be achieved.

What: based, on the one hand, on our experience of transboundary water situations, for one author in the Balkans, for the other author in the European Union and in the Middle East, and, on the other hand, on our work on the UNESCO project of Transboundary Groundwater Education and Training (Fried and Ganoulis (eds) 2016), our approach for an effective and "good" transboundary hydro-governance is based on the integration of scientific, technical, economic, sociological and political drivers exposed previously, in a multidisciplinary approach, accounting for both surface water and groundwater as well as water quality and quantity, at four different scales: local, regional, national and international, with the participation of various stakeholders including the public.

How: "Good" shared hydro-governance is suggested to be achieved by using seven progressive stages or social drivers, four of which are active at domestic and three at international level (Fig. 8.1). They can be summarized as follows:

Fig. 8.1 "Good" shared hydro-governance by using seven social drivers

At domestic level:

1. Establish, use and improve a solid and reliable *scientific and technical basis*, integrating, at river basin scale, surface and groundwater management, data collection and information sharing (see Sects. 4.3.1, 6.1, and 6.3)
2. Develop, maintain and optimise a fair *legal and policy framework* by revitalising the consultative decision-making cyclical process as analysed in Sects. 4.1.2, 5.2.1, and 6.4
3. Promote *effective participation* at domestic level and *fund-rising* through an accountable, efficient and responsible water resources management plan (Chap. 5)
4. *Integrate* water resources management *at different scales* (national, regional, local) by distributing different tasks to various organisations in a coordinated way (Sect. 4.2)

At international level:

5. Use effective "*water diplomacy*" to establish cooperation between riparian countries and conclude bilateral cooperation agreements (Sect. 6.5)
6. Institutionalize the relationship between riparian countries, most usually through a Transboundary River Basin Organization (TRBO), sometimes called Transboundary Basin Organizations (TBO) or International River Basin Organization (IRBO), in order to implement joint water strategies and joint water management plans (Sect. 5.3)
7. Develop and improve *education and professional training* for capacity-building, with an emphasis on joint international short intensive courses, and an effort at overcoming cultural and language barriers.

We shall now critically examine these seven social drivers, to identify how they could be built, how they could be coordinated and how they could operate. Our purpose is to yield what we propose as a practical analytical model of "*good*" *transboundary hydro-governance* and its mode of operation, considering that Transboundary Hydro-Governance is one component of a more comprehensive political framework made up of sectoral economic policies, such as agriculture, industry, tourism, energy or transport among others, where water play a significant role, but with their own political and economic priorities, which may create conflicting situations, and influence negotiations in the water domain.

8.2 "Good" Transboundary Hydro-Governance in Seven Stages

8.2.1 Stage 1: Science and Technique

The first stage concerns the shared water resource itself, surface water basin or/and aquifer systems, identifying the parameters which enable to know the resource in all its dimensions, evaluating what data are known and by whom and what data should be completed and how often. This stage's function is to improve the scientific and technical knowledges of the shared water system. First it is needed to clearly define which geographical and hydrological parts of it belong to which country and/or influence each country: geographical delimitation of shared water bodies, typology of transboundary water bodies, ecological status and related pressures, protected areas, monitoring networks.

During this stage, besides the acquisition of knowledge, an important feature is to disseminate this knowledge among the professionals, of course, but also among the stakeholders, to raise the awareness of the populations and their political decision-makers and technical staff on these scientific issues. This implies the organization of communication actions such as field and laboratory public visits, public lectures, the elaboration and diffusion of information leaflets and documents, and, in general, the involvement of the water authorities existing on both sides of the boundary in the information of the public.

To summarize, the objectives of this first stage are to evaluate the status of the scientific and technical knowledge and survey the issues, to process the scientific information and elaborate the components of science-informed decision-making as well as science-informed public participation in hydro-governance.

Science, Technique and their issues constitute the **first driver** of transboundary hydro-governance.

8.2.2 Stage 2: Policy and Law

The second stage is political, to raise public and decision-makers awareness on both the challenges of sharing water and the existing legislation in each concerned country. As we mentioned in Sect. 5.2.1 for the driver "Legislation and Policy", the laws and regulations are prepared, presented and adopted according to the institutional and legal organization of the concerned countries, and may be completed by customary laws and existing rights. Therefore, raising the awareness of the citizens of one country for the laws of the partner country or countries is not a simple matter and, in practice, could be the task of the TRBO mentioned above, which could be charged with the administration, performance monitoring and eventual enforcement of the joint policies and action-oriented management plans, to be completed by awareness-raising actions on both sides of the boundary.

Awareness-raising in policy and law across the boundaries of the riparian countries constitute the **second driver** of transboundary hydro-governance.

8.2.3 Stage 3: Socio-economics

Economic considerations are required for sound water governance and management, hence it is necessary to evaluate what economic constraints affect the water system on each side of the political boundary, what kind of stakeholders are concerned by the shared water system and what are the economic and social issues that affect the shared waters and their prioritization. This can be quite delicate because the state of development and the economies of the riparian countries can differ, involving differences in priorities that may not be always compatible. This is where the participation of stakeholders, including the general public, becomes most significant, through their social distribution and the nature of their involvement, the way they influence policy and management, or political economy, often with conflicting interests whether they belong to a given country or are located on both sides of a political and/or administrative divide. Of course, it also means that the public is correctly informed, especially science-informed, which is an objective of stage 1. Finally, it is Civil Society in the riparian countries, which should play a significant role in transboundary hydro-governance through informed **participation** in transboundary hydro-governance.

Another aspect is fundamental, whether in a given country or in a transboundary situation: **funding.** It is necessary to have access to adequate funding from the countries sharing the water system and possible external donors. The funds should preferably be specifically dedicated to and managed by the Transboundary Basin Organization.

Public informed participation in decision-making, investment in appropriate infrastructures and the funding of joint policies and strategies constitute the **third driver** of transboundary hydro-governance.

8.2.4 Stage 4: Scaling

As introduced in Sects. 4.2.2 and 4.2.3, different authorities and administrations are responsible for water management problems at different levels, such as the River Basin, the local administrative level, the regional level and the central national level (Government and Ministries). These authorities take different forms from country to country and their legal status (public, private or mixed) depends on the particular country's regime (federation, one central or regional governments). Hydro-governance is related to the institutional levels of decision-making, local, regional, international, and the relationships between these levels expressed as relationships between geographic units and political units of all sizes constitute the scale problems.

The coordination between the different authorities and administrations responsible for water management at these different levels is part of hydro-governance and the problems of scale must be clarified to understand how and at which level an efficient hydro-governance can be defined and implemented, and how the various levels are linked and coordinated.

Scaling is the **fourth driver** of hydro-governance.

8.2.5 *Stage 5: Diplomacy and Cooperation*

This stage consists in identifying the demands concerning water of the riparian countries that could generate competition, to stimulate cooperation and to establish specific relationships between these countries to find solutions to competing demands and solve possible water conflicts, and, more generally, to manage their shared waters efficiently for the greater benefit of the concerned populations. We will show that a solution to such an issue is to direct the cooperation between the riparian countries towards the establishment of Transboundary River Basin Organizations (TRBOs) or Transboundary Basin Organizations (TBOs). More generally, stage 5 presents a governance problem, consisting in identifying the political relationships between the concerned countries, especially the existence of a cooperation on non-water matters, such as commerce, food or energy among others, and then conceiving, designing and building the institutional structures and their legislative and regulatory frameworks that should facilitate cooperation between the respective entities in charge of managing water in each country. We have to move from a state-centric approach to a greater involvement of non-state actors influencing state practice through a variety of mechanisms: according to Blatter and Ingram (2000), although states play significant roles in domestic politics and the international arena, they are "neither the only powerful hierarchical actors in domestic politics nor the sole representatives of a monolithic national interest in international politics", hence, the main actors in transboundary water policies are "collective and corporate actors, including agencies from different sectors and levels of government, non-governmental organizations, corporations, and scientific communities". The scattering of responsibilities is probably not the best way to efficiently manage transboundary waters for the greatest benefit of the riparian countries or entities. And the examples, which essentially concern the management of international river basins and which we examined in Chap. 7, confirm that a Transboundary Basin Organization, TBO, or Transboundary River Basin Organization, TRBO, should be the right institutional answer. Moreover, as stated by S. Schmeier (Schmeier 2010), "it is argued that water itself creates interdependencies among states sharing a transboundary watercourse which often goes as far as creating joint institutions for the management of the shared resource, themselves acting as a trigger for further cooperation". "Once cooperative water regimes are established, they turn out to be tremendously resilient over time, even between otherwise hostile riparians and even as conflicts is waged over other issues" (Wolf 1998, p. 194). This holds true even in times of derogating relations between involved states, which could be shown by Kalpakian (2004) for the case of India and Pakistan, where cooperation based on the Indus Treaty prevailed despite the Indo-Pakistani wars, and for the Mekong River Basin, where cooperation in form of the Mekong Committee and the Interim Mekong Committee prevailed despite ongoing ideological confrontations and the Indochinese Wars". (Schmeier 2010; Kalpakian 2004; Wolf 1998).

But such an institution will efficiently ensure cooperation between the various governmental entities on both sides of the boundary if it has a strong enforcement capacity resulting from a strong mandate and political support from the concerned governments involving a political agreement between the riparian countries achieved through diplomatic mechanisms, within a consolidated legal framework that provides the capacity to develop policies and laws and allows for the adoption of decisions and their implementation completed by mechanisms of reporting; if it disposes of adequate funding to support joint programs and structures; if it involves the participation of the various stakeholders, e.g. private and public water companies and water management agencies at national and local levels, users' associations, donors, research institutions and the general public. It should provide the stakeholders the possibility to share their concerns, propose their solutions and, by offering them a platform of interaction, discussions and exchanges of views, contribute to develop trust and confidence. (*UN-Water Thematic Paper: Transboundary Waters, sharing benefits, sharing responsibilities* 2008) (*the Handbook for Integrated Water Resources Management in Transboundary Basins of Rivers, Lakes, and Aquifers INBO, GWP,* 2012). In Chap. 7, we have presented successful examples of such Transboundary Basins Organizations (TBOs), also called International River Basin Organizations (IRBOs, or simply RBOs or River Basin Organizations) or Transboundary River Basin Organizations (TRBOs), but also noticed that the existing TBOs do not usually show all the characteristics mentioned above. In stage 6, we will examine the types, structures and operations of Transboundary River Basin Organizations.

From these considerations, we conclude that Cooperation and Diplomacy constitute the **fifth driver** of transboundary hydro-governance.

8.2.6 Stage 6: Joint Management/RBO

Water is directly connected to different economic activities, such as food, energy, industry, tourism and is the basis for environmental protection and ecosystem preservation and, through its governance, water can facilitate the development of all these economic sectors. Historically, the example of the Tennessee Valley Authority (TVA) in the USA, although operating in a non-international context, is frequently cited as the prototype institution for enhancing not only successful water resources governance between different states, but also for stimulating sustainable socio-economic development in the region. In fact, water is directly connected to different economic activities, such as food, energy, industry, tourism and is the basis for environmental protection and ecosystem preservation and, through its governance, water can facilitate the development of all these economic sectors. From the analysis of the examples we have presented in Chap. 7 and from our own experience, the creation of international joint institutions such as a Transboundary River Basin Organisation (TRBO) seems to be a most efficient governance instrument for implementing effective transboundary hydro-governance.

Hence, we can only recommend such creation, for instance noticing that the establishment of RBOs at national level is the main institutional tool introduced by the EU-WFD (European Union Water Framework Directive) to apply the IWRM process. We remind the reader that, in Chap. 5, the structure, legal base, decision-making processes and other conditions have been examined to create effective international institutions, which, according to their level of specificity, could be Advisory Committees, Authorities with some planning powers or International Commissions.

Stage 6 consists in choosing the type of Transboundary Basin Organization and how it operates. We propose that the choice be made among the four institutional types identified by Hooper and Lloyd in their 2011 Report on IWRM in Transboundary Basins, which are realistic, and which we introduced in Chap. 5: an advisory committee, an association (or initiative), an authority, or a commission.

Anyway, whatever its type, a Transboundary Basin Organization should be structured by a permanent secretariat with

- scientific responsibilities, such as the coordination of monitoring activities and data processing or developing joint monitoring programs,
- policy responsibilities, such as the formulation of joint policies and their implementation,
- capacity-building responsibilities, such as the pooling of existing capacities in riparian countries,
- financial responsibilities, such as raising and managing financial support,
- arbitration responsibilities, such as the settlement of disputes or differences, and, in general,
- coordination functions and the monitoring and reporting of policy implementation.

A Transboundary Basin Organization should operate along appropriate rules of procedure and terms of reference that consider specific local conditions. These rules should not only recommend the structure, responsibilities, rights and financial status of such an institution, but also ways and means to ensure public participation, and, more generally, the informed participation of all stakeholders. Schmeier (2010) has rather thoroughly presented the mechanisms of River Basin Organizations, namely their decision-making processes, the corresponding needs for data and information sharing within the River Basin Organization, the dispute resolution mechanisms, and, finally, the very essential financial resources and their possible sources. As we state in Sect. 8.2.5, TBOs do play a role in the design and implementation of joint water policies and strategies by assisting the riparian countries authorities in their negotiations with their logistics and also as facilitators because of their knowledge of the transboundary situation in both the water and the non-water domains.

Furthermore, a Transboundary Basin Organization will rely on competent human capacities: the staff should have broad competences and skills, be multidisciplinary and able to coordinate to achieve an integrated view of a problem and effectively solve it in the real world. They should be able to cover the classical requirements of water governance and management as described in Chap. 5, such as, for instance, implementing policies and laws or setting up funding arrangements or performing IWRM, and they should also be able to perform the various tasks of negotiation, diplomacy and conflict resolution. It means that capacity-building will play an important role, as presented in stage 7.

It clearly appears that Joint Management and the setting of Transboundary River Basin Organizations constitute the **sixth driver** of transboundary hydro-governance.

8.2.7 Stage 7: Capacity-Building

Stage 7 will consist in education and professional training for capacity-building, to know how to establish and operate a Transboundary River Basin Organization in a multidisciplinary approach. It is necessary to conceive and design curricula at two levels of action, a policy level and a practitioner's level:

At policy level, the aim is to create the capacities to have a very general vision of the transboundary water problems, including cultural and language differences and international political and economic issues, to take part in negotiations, to make political and managerial decisions, and to create the institutions and instruments necessary for transboundary management, especially joint institutions and partnerships.

At practitioner's level, the aim is to train those who, in the field, will work together to consider the different national legal systems, the international political and economic issues and the national priorities, and will, then, implement the political, institutional and managerial decisions, scientifically, legally and economically.

Such a training programme will follow a multidisciplinary approach involving hydrogeology, water management, economy, law, engineering, political science, and will put an emphasis on the communication dimensions between the different future actors of transboundary water policy and management, which may necessitate a common language, both linguistic and scientific, to facilitate the teaching of these future actors together. To illustrate what type of topics we think to be adequate, here is a shopping list of the proposed topics that we identified during the preparatory seminar "Designing a UNESCO Training Programme in Transboundary Groundwater Management" held in Paris, followed by two pilot courses respectively held in October 2008 in Thessaloniki (Greece) and in December 2010 at UNESCO Headquarters in Paris (France):

- Transboundary groundwater concepts, including monitoring and modelling.
- Environment: ecosystems, biodiversity, climate change, human activities including contamination.
- Legal, institutional and policy framework (including state-of-the-art).
- International law between countries, human rights law, domestic comparative laws, transfrontier laws between local authorities.
- Institutional capacity building.
- Joint mechanisms for institutional coordination.
- Institutional and political mechanisms for decision-making

- Communication between policy-makers, decision-makers and scientists (bottom-up and top-down approaches)
- Economic and social aspects.
- Public participation: communication, networking, human perceptions.
- Transboundary management instruments for cooperation.
- Integrated Water Resources Management (IWRM).

To summarize, the general philosophy of the capacity-building we deem necessary to support transboundary water governance is the following: while the policy level should be the domain of the social sciences and humanities with knowledge of water resources and management, the practitioners' level should generally be the domain of environmental scientists, engineers, hydrologists and those trained in the physical understanding of resource management. It will be important to create linkages between policy and science so that there is common understanding and the ability to communicate across regional and international boundaries. For example, language, history, cultural practices and traditions are important elements for bargaining and mutual decision-making.

Specific short intensive courses should be made available, on the example of these UNESCO pilot courses and the publication of a Manual (Fried and Ganoulis (eds) 2016). Of course, a survey of existing training capacities in the riparian countries should be performed, and, if adequate, their pooling by the Transboundary Basin Organization should be proposed.

This development of skills and talent constitutes the **seventh driver** of transboundary hydro-governance.

References

Blatter J, Ingram H (2000) States, markets and beyond: governance of transboundary water resources. Nat Resources J 40:439. http://digitalrepositoryunmedu/nrj/vol40/iss2/12. Accessed 27 Dec 2017

Fried J, Ganoulis J (eds) (2016) Transboundary groundwater resources-sustainable management and conflict resolution. Lambert Academic Publishing, Saarbrücken

Kalpakian J (2004) Identity, conflict and cooperation in international river systems. Ashgate Publishers, Aldershot

Schmeier S (2010) Governing international watercourses. . .Hertie School of governance. Working papers 53, Aug 2010

The Handbook for Integrated Water Resources Management in Transboundary Basins of Rivers, Lakes and Aquifers (2012) Global Water Partnership/International Network of Basin Organizations, Stockholm

UN Water Thematic Paper Transboundary Waters (2008) Sharing benefits, sharing responsibilities

Wolf A (1998) Conflict and cooperation along international waterways. Water Policy 1:251–265

Conclusion

We did suggest an analytical model for "good" transboundary hydro-governance, based on our own experience and the interpretation of a series of case histories worldwide that we briefly presented in this book.

Although we are rather optimistic through our personal experience in the field and think that our model is a step forward for understanding and implement effectively a practical governance of transboundary waters, we also share the following views expressed by Earle and Neal (2017):

"Despite the existence of international- and basin-level agreements and basin organizations, the benefits to be expected from international cooperation around transboundary watercourses have in most cases not materialized. Acute, persistent and seemingly intractable problems persist, with ecosystem degradation not being reversed, joint investments in water infrastructure not materializing and joint management organizations failing to attract significant long-term support from the respective basin states. Despite at least two decades of concerted support by the international development community, the impacts of enhanced interstate cooperation are noticeable through their absence".

We add that only a few cases of successfully managed transboundary aquifers do exist and have been presented in this book. There still is a long way to go!

Reference

Earle A, Neal MJ (2017) Inclusive transboundary water governance. In: Freshwater governance for the 21st century, chapter 8, Global issues in water policy 6, p 145–148

J. Ganoulis, J. Fried, *Transboundary Hydro-Governance*,
https://doi.org/10.1007/978-3-319-78625-4

The manufacturer's authorised representative in the EU is Springer Nature Customer Service Centre GmbH, Europaplatz 3, 69115 Heidelberg, Germany. If you have any concerns regarding our products, please contact ProductSafety@springernature.com

Printed and bound by CPI Group (UK) Ltd, Croydon, CR0 4YY
30/09/2026
02225654-0006